THE CHEMISTRY OF BERYLLIUM

TOPICS IN INORGANIC AND GENERAL CHEMISTRY

A COLLECTION OF MONOGRAPHS EDITED BY

P. L. ROBINSON

*Emeritus Professor of Inorganic Chemistry
in the University of Durham,
King's College, Newcastle upon Tyne*

VOLUME 1

ELSEVIER PUBLISHING COMPANY
AMSTERDAM / LONDON / NEW YORK

THE CHEMISTRY OF BERYLLIUM

BY

D. A. EVEREST, Ph.D., F.R.I.C., A.M.I.M.M.

*The National Chemical Laboratory,
Teddington (Great Britain)*

ELSEVIER PUBLISHING COMPANY
AMSTERDAM / LONDON / NEW YORK
1964

SOLE DISTRIBUTORS FOR THE UNITED STATES AND CANADA
AMERICAN ELSEVIER PUBLISHING COMPANY, INC.
52 VANDERBILT AVENUE, NEW YORK 17, N.Y.

SOLE DISTRIBUTORS FOR GREAT BRITAIN
ELSEVIER PUBLISHING COMPANY LIMITED
12B, RIPPLESIDE COMMERCIAL ESTATE
RIPPLE ROAD, BARKING, ESSEX

LIBRARY OF CONGRESS CATALOG CARD NUMBER 63-16078

WITH 5 ILLUSTRATIONS AND 6 TABLES

ALL RIGHTS RESERVED
THIS BOOK OR ANY PART THEREOF MAY NOT BE REPRODUCED IN
ANY FORM, INCLUDING PHOTOSTATIC OR MICROFILM FORM,
WITHOUT WRITTEN PERMISSION FROM THE PUBLISHERS

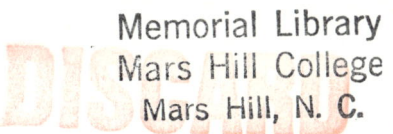

Preface

The object of the present book has been to provide a concise but essentially complete account of the chemistry of beryllium. The present heightened interest in beryllium metal and its compounds, especially with regard to their applications in the fields of atomic energy and space technology, is reflected in the considerable increase in the volume of work published on the chemistry of beryllium in the last few years. Such considerations, coupled with the fact that there has been no recent general account of beryllium chemistry, indicates that the present work comes at an opportune time.

Discussion of beryllium metallurgy has been deliberately omitted from this monograph as this subject has been dealt with adequately elsewhere. Specialised applied topics have also been omitted, *e.g.* applications of beryllia refractories, except where they bear directly upon general beryllium chemistry. However, a discussion of the extractive metallurgy of beryllium has been included as this subject forms an integral part of beryllium chemistry.

I would like to thank Miss E. Napier, Mr. R. A. Wells and my other colleagues at the National Chemical Laboratory for their generous help and advice. In particular, I would like to express my thanks to Professor P. L. Robinson for the kindly and unstinting help and encouragement he has given to me throughout the writing of this monograph. My thanks also go to Mrs. J. Chalk for typing my manuscript.

Teddington,
September 1963
D. A. EVEREST

Contents

Preface. v

Chapter 1. *Introduction to Beryllium Chemistry*

1. History 1
2. General chemical character of beryllium 2
3. Valency and charge state of beryllium 3
4. Beryllium of charge number one 6
   References . . 6

Chapter 2. *Solution Chemistry of the Simple Be^{2+} Ion*

1. General considerations 7
2. Solution chemistry of the hydrated Be^{2+} ion in weakly acidic media 8
3. Beryllium hydroxide 12
4. Acidic reactions of beryllium hydroxide 14
5. Interactions of beryllium cations with ion-exchange resins and solvents 16
6. Beryllium in liquid ammonia 19
   References . . 21

Chapter 3. *Simple Oxosalts of Beryllium*

1. Beryllium carbonates. 23
2. Beryllium sulphate 25
3. Beryllium selenate 27
4. Beryllium nitrate. 28
5. Beryllium phosphates. 29
6. Beryllium arsenates 30
7. Beryllium oxalate 30
8. Normal beryllium salts of carboxylic acids 32
9. Beryllium perchlorate 33
10. Beryllium azide 33
11. Beryllium iodate and periodate 33
12. Beryllium sulphonates 33
13. Beryllium silicates 34
14. Beryllium cyanide 35
    References . . 35

Chapter 4. The Beryllium Halides

1. Preparation of beryllium fluoride 38
2. Properties of beryllium fluoride 39
3. Aqueous solutions of beryllium fluoride 42
4. Fluoroberyllates prepared from aqueous solution 43
5. Anhydrous beryllium fluoride systems 45
6. Beryllium fluoride glasses 48
7. Miscellaneous reactions of beryllium fluoride 49
8. Preparation of beryllium chloride. 50
9. Properties of beryllium chloride 50
10. Aqueous solutions of beryllium chloride 52
11. Anhydrous beryllium chloride systems 53
12. Beryllium chloride complexes with neutral ligands 54
13. Miscellaneous reactions of beryllium chloride 56
14. Preparation and properties of beryllium bromide 57
15. Reactions of beryllium bromide 57
16. Preparation and properties of beryllium iodide 58
17. Reactions of beryllium iodide 59
   References . . 59

Chapter 5. Complex Beryllium Compounds

1. Factors governing the stability of beryllium complexes 64
2. Beryllium complexes with 1,3-diketones 67
3. Beryllium complexes with hydroxy acids 69
4. Beryllium oxide carboxylates (basic carboxylates) 71
5. Beryllium co-ordination polymers 76
6. Beryllium-phthalocyanine complexes 78
   References . . 79

Chapter 6. Simple Binary Compounds of Beryllium

1. Beryllium hydride 82
2. Beryllium borides 84
3. Beryllium carbide 84
4. Beryllium nitride. 85
5. Beryllium oxide 86
6. Beryllium sulphide 88
7. Beryllium selenide and telluride 88
8. Beryllides 88
   References . . 89

Chapter 7. Organo-beryllium Compounds

1. Dimethylberyllium 91
2. Diethylberyllium. 95
3. Di-*iso*propylberyllium 96
4. Di-*tert*.-butylberyllium 97
5. Diphenylberyllium 97

CONTENTS

6. Dicyclopentadienylberyllium 98
7. Catalytic reactions 98
8. Beryllium compounds of the spiran type.......... 99
9. Beryllium alcoholates and phenolates 99
   References . . 100

Chapter 8. *The Extractive Metallurgy of Beryllium*

1. Beryllium ores...................... 102
2. Extraction of beryllium from beryl 104
3. Copaux fluorosilicate process 104
4. Sulphate routes for recovery of beryllium from beryl 109
5. Chlorination methods 113
6. Preparation of metallic beryllium 113
7. Magnesium reduction of beryllium fluoride 114
8. Electrolysis of beryllium chloride 115
   References . . 116

Chapter 9. *The Analytical Chemistry of Beryllium*

1. Introduction 117
2. Breakdown of beryllium-containing samples 117
3. Quantitative procedures for beryllium determination 118
4. Newer methods for beryllium separation 123
5. Radiochemical separation procedures 125
6. Radiometric methods of beryllium analysis 126
7. Spectrometric methods 129
8. Polarographic methods 130
   References . . 131

Chapter 10. *The Beryllium Health Hazard and its Control*

1. An outline of the occurrence and nature of beryllium disease . 133
2. Health control in a beryllium laboratory 135
3. Maximum permissible concentrations 138
   References . . 140

Chapter 11. *Nuclear Properties and Reactions of Beryllium*..... 141
   References . . 145

Subject Index....................... 146

CHAPTER 1

Introduction to Beryllium Chemistry

1. History

Although beryllium compounds have been known in the gemstone forms of beryl, emerald and aquamarine, since before 2000 B.C., it was not until 1798 that Vauquelin isolated from beryl the new oxide, "*la terre du beril*"[1]. The oxide differed from alumina in furnishing salts with a sweet taste, in not forming an alum, in being soluble in ammonium carbonate and in not being precipitated by potassium oxalate or tartrate. It was at the suggestion of the Editors of Annales de Chimie, clearly impressed by the singular taste of the salts, that Vauquelin referred to this compound as glucine. In Germany the name berylerde, obviously a translation of *la terre du beril*, was preferred; it was not until 1828 that the name beryllium was used for the element[2]. This name is now always employed except in French chemical literature, where glucinium (Gl) is still used.

The valency, or as we should now say charge number in referring to an essentially electrovalent condition, of beryllium was the subject of controversy for many years. Berzelius first regarded beryllium as possessing a charge of either two or four and formulated the oxide as either BeO or BeO_2, but in 1815 he was constrained by its strong chemical resemblance to aluminium to change this to three. A combining ratio of beryllium of 4.7 was not in dispute and thus 14 was accepted as the atomic weight. This view appeared to receive support from the application of the Dulong and Petit rule [atomic weight = 6.3/specific heat of beryllium (0.397) = 15.8] and from Mitscherlich's law, as beryllium was then considered able to replace aluminium in such silicates as $K_2O.Al_2O_3.6SiO_2$. But

References p. 6

as Mendeleev was unable to accommodate an element of atomic weight 14 in his periodic table, with characteristic prescience he maintained that the chloride was $BeCl_2$ and the atomic weight 9. It thus fell comfortably between lithium and boron and there he put it in his 1869 table.

Mendeleev was proved correct by the accurate vapour density measurements on beryllium chloride of Nilson and Patterson[3]. By using sufficiently dry and pure material they avoided the difficulties due to traces of moisture which had vitiated the results of earlier workers; their values were:

T°	589	597	604	686	720	745	812
	3.067	3.031	3.090	2.853	2.926	2.753	2.793

Stoichiometric $BeCl_2$ requires 2.77. These results were fully supported by the later work of Humpidge[4] who obtained 2.733 at 635° and 2.714 at 785°.

2. General chemical character of beryllium

Beryllium, with an atomic number of four and an atomic weight of 9.013 (chemical scale)[5], lies in the first short period of the periodic table; it heads Group IIA which includes magnesium, calcium, strontium, barium and radium. The electronic structure, $1s^2\,2s^2$, has an outer shell of two s electrons which is characteristic of the elements of this family. Below this ns^2 shell every member has a complete inert gas core of electrons. Thus the beryllium ion, and the other alkaline earth ions, all possess the structure of the inert gas of atomic number lower by two units and a bipositive charge.

Table 1 shows properties of beryllium in relation to those of the other alkaline earth elements. Of these the low ionic radius and the high melting and boiling points are noteworthy. At room temperature, the metal has a close-packed hexagonal structure similar to magnesium, the lattice parameters at 20° are: a = 2.286, c = 3.583 Å, c/a = 1.568. It has a steel grey colour and resembles

TABLE 1
PROPERTIES OF THE GROUP IIA ELEMENTS

	Be	Mg	Ca	Sr	Ba	Ra
Density g/cm^3	1.85	1.74	1.55	2.6	3.78	5.0
Atomic volume cm^3/g	4.85	14.0	26.3	33.6	36.4	45.2
Ionic radius Å	0.31	0.65	0.99	1.13	1.35	
M.p. °C	1283	651	851	800	850	960
B.p. °C	1500 (at 5 mm)	1170	1437	1366	1537	1140

aluminium in resisting oxidation in spite of its oxide having a heat of formation ($-\Delta H_{298}°$) of 143.1 kcal/mol. This is a kinetic effect arising from the volume of beryllium oxide being larger than that of the metal from which it is formed so that it produces a surface film which protects the metal from further attack. Cold concentrated nitric acid passivates beryllium but the metal is dissolved by hydrochloric acid and dilute sulphuric and nitric acids. Beryllium metal is dissolved by an aqueous solution of sodium or potassium hydroxide with the evolution of hydrogen and formation of a beryllate. It is also rapidly dissolved by aqueous ammonium bifluoride:

$$2NH_4HF_2 + Be = (NH_4)_2BeF_4 + H_2$$

This reaction is employed for the recovery of scrap beryllium (see Chapter 8).

3. Valency and charge state of beryllium

Although usually considered as ionic, the bonds formed by beryllium in many compounds have considerable covalent character, as indeed would be expected from the high electronegativity[6] of beryllium (1.5). In the 1s^2 2s^2 state beryllium, like helium, is zerovalent, but it can form covalent bonds by the promotion of one 2s electron to the 2p level which results in two linear hybrid sp bonds. The energy difference between these levels is only 62 kcal., very much less than the 460 kcal. needed to promote an electron in the helium atom from the 1s to the 2s state. The gaseous monomeric beryllium halides are examples of such linear molecules. In a number of instances apparently simple covalent beryllium compounds are actually polymeric. These compounds are classed as electron deficient, since there are insufficient electrons present to form the required number of two electron bonds. Beryllium hydride (Chapter 6) and the beryllium alkyls (Chapter 7) are examples of such electron deficient molecules.

Beryllium forms a wide range of complex compounds in which it accepts a share in two extra pairs of electrons to form four tetrahedral sp^3 hybrid bonds.

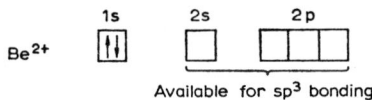

$$\overset{\delta+\quad\delta-}{}$$

Such bonds have considerable polarity Be—X, and beryllium complexes have an ionic character (*cf.* Chapter 5). Unlike the other alkaline earth elements and zinc, beryllium cannot expand its co-ordination number to six. This is due to two causes: the small size of the beryllium atom and the inaccessibility of the 3d orbitals which are required for sp^3d^2 hybridisation. It is a feature of all the elements in the first short period of the periodic table.

Although anhydrous beryllium compounds are at best only partly ionic, the Be^{2+} ion certainly occurs in hydrated form in solution (Chapter 2) and in hydrated beryllium salts (Chapter 3). As it has a crystal radius[6] of only 0.31Å, the Be^{2+} ion possesses the very high charge to radius ratio (z/r) of 6.45. This high value would itself indicate that beryllium compounds possess marked covalent character. The value of 6.45 may be compared with the z/r ratios of 1.67, 6.00, 3.07 and 2.02 for the Li^+, Al^{3+}, Mg^{2+} and Ca^{2+} ions respectively. These values reflect both the chemical similarity of beryllium and aluminium and the sharp difference between the properties of beryllium and those of the other alkaline earth elements.

The difference between beryllium and the other alkaline earth elements is also reflected in the ionization potentials of the neutral atoms (Table 2). These potentials also bring out the chemical resemblances of beryllium and zinc; there are zinc oxide carboxylates corresponding to the beryllium oxide carboxylates and, amongst minerals willemite, Zn_2SiO_4, is very similar to phenacite, Be_2SiO_4. These strong similarities have even led to the suggestion that beryllium and magnesium should be associated with the zinc rather than the calcium sub-group[7]. This similarity shown by zinc for beryllium is due to the polarizability of the eighteen electron

TABLE 2

IONIZATION POTENTIALS IN ELECTRON VOLTS FOR ELEMENTS IN GROUPS IIA AND IIB OF THE PERIODIC TABLE

	1st	2nd
Be	9.320	18.21
Mg	7.644	15.03
Ca	6.111	11.87
Sr	5.692	10.98
Ba	5.210	9.95
Zn	9.311	17.89
Cd	8.991	16.84
Hg	10.434	18.65

sub-shell of the transition type Zn^{2+} ion ($3s^2\,3p^6\,3d^{10}$) being greater than that of the inert gas type Ca^{2+} ion ($3s^2\,3p^6$). As a result, the total polarization interaction between Zn^{2+} and an anion is nearly as great as the interaction between the anion and the less polarizable but more strongly polarizing Be^{2+} ion, and greater than the interaction between the Ca^{2+} ion and an anion.

It is of interest to compare the 1st ionization potential of beryllium with those of its neighbours in the first short period of the periodic table. In the first place, it is greater than that of lithium (5.39 eV) because, although both electrons are 2s, beryllium has a greater nuclear charge than lithium. It is also greater than that of boron (8.30 eV) because its electron in the 2p orbital is subject to the screening effect of the closed 2s shell. This is sufficiently large to overcome the effect of the increased nuclear charge. With carbon and nitrogen the two extra electrons enter the other two 2p orbitals,

References p. 6

and the ionisation potential increases to 11.285 and 14.545 eV, respectively, because of the increased nuclear charge and an unchanged screening.

4. Beryllium of charge number one

Although beryllium almost invariably possesses a charge number of two, there is evidence that it can display a charge number of one under certain conditions. Thus the electrolysis of a sodium chloride solution between beryllium electrodes in a divided cell produces beryllium of both charge numbers[8]. The Be^+ formed decomposes water with evolution of hydrogen and on prolonged electrolysis decomposition occurs:

$$2Be^+ \rightarrow Be + Be^{2+}$$

Beryllium(I) is also formed during the electrolysis of molten beryllium chloride–sodium chloride mixtures with a beryllium anode (see Chapter 4).

REFERENCES

1 VAUQUELIN, L. N., *Ann. Chim. Phys.*, 26 (1798) 155 et. seq.
2 WÖHLER, F., *Pogg. Ann.*, 13 (1828) 577.
3 NILSON, L. F. AND PATTERSSON, O., *Ber.*, 11 (1878) 906; 17 (1884) 987; *J. prakt. Chem.*, 33 (1885) 1.
4 HUMPIDGE, T. S., *Proc. Roy. Soc.*, 35 (1883) 137, 358; 39 (1886) 1.
5 JOHANNSEN, T., *Naturwissenschaften*, 31 (1943) 592.
6 PAULING, L., *The Nature of the Chemical Bond*, 3rd ed. Cornell University Press, Ithaca, N.Y., 1960.
7 PFEIFFER, P. et al., *Z. Anorg. Chem.*, 264 (1951) 183.
8 LAUGHLIN, B. D., KLEINBERG, J. AND DAVIDSON, A. W., *J. Amer. Chem. Soc.*, 78 (1956) 559.

CHAPTER 2

Solution Chemistry of the Simple Be^{2+} Ion

This chapter deals principally with the solution chemistry of the hydrated Be^{2+} ion in both acid and alkaline media. The chemistry of beryllium hydroxide will also be included as it is considered to be closely related to the subject under discussion. Particular attention will be given to the strong tendency of beryllium to form polynuclear species, a fact which dominates the solution chemistry of this element. In addition, reactions of the Be^{2+} ion with ion-exchange resins and solvents will be considered.

1. General considerations

As in so much of its chemistry, the behaviour of the Be^{2+} ion is governed by its very high charge to radius ratio. An immediate consequence of this is that the Be^{2+} ion is, as Gurney says[1], a strong order-producing ion, implying that it strongly orientates the water molecules in its immediate vicinity, thus bringing them to a greater degree of order than they formerly possessed. This is shown by the low partial molar entropy $\bar{S}^°$ (−55 e.u. for Be^{2+} compared with −28.2, −13.2, −9.4 and 3 e.u. for Mg^{2+}, Ca^{2+}, Sr^{2+}, and Ba^{2+} respectively)[2] and high positive viscosity B coefficient of the Jones and Dole equation[3] which is possessed by the Be^{2+} ion. That the ion strongly orientates water molecules in solution, *i.e.* it is heavily hydrated, has been known for many years. For example, Fricke[4] in 1923, from considerations of the high viscosity of beryllium salt solutions and the small mobility of the Be^{2+} ion and from freezing point measurements, concluded that the Be^{2+} ion was the most heavily hydrated of all the bivalent ions. Later values for the hydration of the Be^{2+} ion have been summarized[5].

This tendency for the Be^{2+} ion to hydrolyse derives directly from its high charge to radius ratio. The resulting high charge density on the surface of this ion both polarizes the surrounding water molecules and causes them to orient with the negative end of their dipoles directed towards the beryllium ion. In effect, the protons of the water molecules are so strongly repelled that, sooner or later, thermal agitation will be sufficient to transfer a proton to a more distant water molecule, thus leaving an OH^- ion in contact with the Be^{2+} ion. The whole process, which obviously increases in extent as the pH of the solution is increased, can be represented:

$$nBe^{2+} + nH_2O \rightleftharpoons (BeOH)_n^{n+} + nH^+$$

The amphoteric properties of beryllium can also be accounted for on a similar basis[6]. Cartledge[7] defines the ionic potential of an ion Φ, as $\Phi = z/r$, where z is its charge and r its radius. He considers that ions with $\sqrt{\Phi} \langle 2.2$ are basic, those with $3.2 \rangle \sqrt{\Phi} \rangle 2.2$ are amphoteric, and those with $\sqrt{\Phi} \rangle 3.2$ are acidic. As $\sqrt{\Phi}$ for Be^{2+} is 2.54 whilst for Mg^{2+} it is 1.76 the fact that beryllium is amphoteric, and the other alkaline earth elements basic, is readily explained. It should be pointed out however, that the above generalization has limits of application, for example barium can be amphoteric under certain conditions[8].

2. Solution chemistry of the hydrated Be^{2+} ion in weakly acidic media

The most important feature of the hydrated Be^{2+} ion is its ready hydrolysis to give polynuclear species. An outstanding effect of this tendency is that beryllium salt solutions dissolve up to several mole proportions of beryllium oxide[9]. Sidgwick and Lewis[10] suggested that the complex ion $[Be(OBe)_x]^{2+}$ is then formed. However, this reaction is best considered as a particular aspect of the normal hydrolytic reactions of the hydrated Be^{2+} ion, which are intensified by the beryllium oxide increasing both the pH and the beryllium concentration of the solution. As will be seen, both

factors increase the hydrolytic polymerisation of the Be^{2+} ion.
The hydrolysis of the Be^{2+} ion has been frequently studied[11-17]. Most workers agree that addition of alkali to an aqueous beryllium salt leaves the solution clear up to $Z = 1$, where Z is the average number of OH^- bound per Be^{2+}, above that value precipitation begins. With halide salts precipitation is complete at $Z = 2$, but with oxoacid salts, such as sulphate or nitrate, it is complete at $Z = 1.8$-1.9. This indicates that the precipitate is not pure $Be(OH)_2$ but contains some of the salt anion. That precipitation occurs only when $Z \rangle 1.0$ suggests the formula $Be_n(OH)_n^{n+}$ for the soluble hydrolytic complex. Prytz[11] favoured a complex with $n = 2$, French

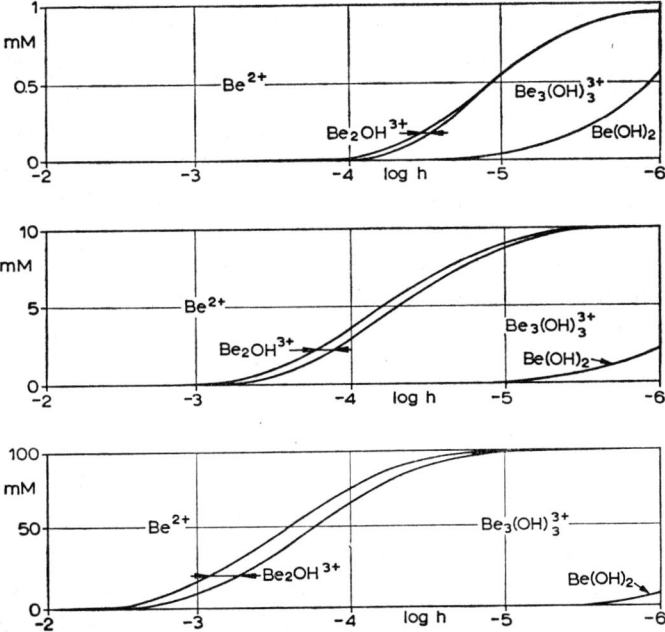

Fig. 1. (1, 2, and 3) Distribution of complexes for total beryllium concentration $B = 1$, 10 and 100 mM, as a function of log h. If a vertical line is drawn for a given value of h, the segment of this line falling in a certain area e.g. Be_2OH^{3+}, represents the concentration of Be present as this complex in the solution at equilibrium.

(Reproduced, with permission, from *Acta Chem. Scand.*, 10 (1956) 1002.)

References p. 21

workers[12-14] a complex with n = 4, and Mattock[15] held that many different ionic species of increasing polynuclear complexity exist in the solutions. At present the generally favoured value is n = 3 suggested by Garrett[16] and, in particular, by Kakihana and Sillén[17] who employed the well known experimental and computational procedures developed by the Stockholm school.

The following equilibria and equilibrium constants are given by Kakihana and Sillén:

$$3Be^{2+} + 3H_2O \rightleftharpoons Be_3(OH)_3^{3+} + 3H^+ \qquad \log \beta^*_{33} = -8.66$$

$$2Be^{2+} + H_2O \rightleftharpoons Be_2OH^{3+} + H^+ \qquad \log \beta^*_{12} = -3.24$$

$$Be^{2+} + 2H_2O \rightleftharpoons Be(OH)_2 + 2H^+ \qquad \log \beta^*_{2} = -10.9$$

The symbols employed are those given in the Tables of Stability Constants[19] (see also footnote on p. 148 of ref. 18). Fig. 1 shows the distribution of the different beryllium species as a function of $\log h$ ($h = [H^+]$) for beryllium concentrations of 1–100 mM. It is seen that $Be_3(OH)_3^{3+}$ is the most important polymer, followed closely by Be_2OH^{3+}, and these polymers exist over a wider pH range as the beryllium concentration in solution increases. Kakihana and Sillén[17] suggest a structure for the $Be_3(OH)_3^{3+}$ complex, in which beryllium atoms are linked by hydroxyl bridges, the four coordination of the beryllium being preserved by extra water molecules.

Although the main lines upon which the hydrolytic polymerisation of the Be^{2+} ion takes place are established, it is probable that the picture is still incomplete. For instance, Kakihana and Sillén[17] suggest the formation of higher complexes at Z⟩0.9, this being favoured by an increase in the beryllium concentration. Supporting this suggestion is the fact that a sulphonic acid cation-exchange resin (Zeo-karb 225) takes up 2.92 g. atoms of beryllium per equiv. resin from an 0.5 M beryllium sulphate solution of pH 5.1; such a loading could be feasibly accounted for by assuming the adsorption of the ion $[Be_n(BeO)_{2n}]^{2n+}$ by the resin[20]. Moreover, as Sidgwick[21] pointed out, it is possible to dissolve more beryllium oxide in a beryllium salt solution than would correspond to the formation of $[BeOH]_n^{n+}$.

Recently relaxation measurements on aqueous salt solutions, often employing ultrasonic absorption, have been shown to provide a powerful method for the study of the aqueous chemistry of cationic species. Such measurements on beryllium sulphate solutions have been interpreted in terms of the scheme[22]:

$$Be^{2+} + SO_4^{2-} \rightleftharpoons Be^{2+} \begin{matrix} H & H \\ O & O \\ H & H \end{matrix} SO_4^{2-} \rightleftharpoons Be^{2+} \begin{matrix} H \\ O \\ H \end{matrix} SO_4^{2-} \rightleftharpoons BeSO_4$$

The first two steps are rapid with average mean times (relaxation times) $\tau_1, \tau_2 \geqslant 10^{-9}$ sec. The last step, involving the replacement of a water molecule in the inner hydration sphere of the beryllium by a sulphate ion, proceeds more slowly, and the velocity is specifically controlled by the particular cation. The relaxation time τ_3 for this process is relatively long, 8×10^{-4} sec; so long in fact, that ultrasonic methods cannot be used for beryllium, and a temperature jump method[23] is used instead. The value of τ_3 for beryllium sulphate is much greater than the value of 10^{-6} sec. for magnesium sulphate. The latter value is the more normal for a bipositive metal sulphate.

For most bipositive ions[24] the rate limiting step (as given by τ_3), for the process described as occurring with beryllium sulphate, is independent of the entering ligand. Beryllium however falls into a second group, which includes the cations Fe^{3+} and Al^{3+}, in which the rate is dependent upon the acidity of the entering ligand. Ions in this group also show other rate characteristics due to reactions of the hydrolytic complexes. Thus between pH 3 and 4.5 beryllium sulphate has two additional relaxation effects (τ' and τ''); the time constants of both of these show a clear dependence upon both the hydrogen and beryllium ion concentrations. These two relaxation effects are interpreted by Diebler and Eigen[22] as indicating the following reaction schemes:

(for τ'') $2BeOH^+ \rightleftharpoons \left[Be\begin{smallmatrix}OH\\OH\end{smallmatrix}Be \right]^{2+}$

and $\left[Be\begin{smallmatrix}OH\\OH\end{smallmatrix}Be \right]^{2+} \rightleftharpoons [Be-O-Be]^{2+} + H_2O$

(for τ') $Be^{2+} + SO_4^{2-} \rightleftharpoons [BeSO_4]$

and $[BeSO_4] + BeOH^+ \rightleftharpoons \left[Be\begin{smallmatrix}SO_4\\OH\end{smallmatrix}Be \right]^{2+}$

3. Beryllium hydroxide

Beryllium hydroxide has at least two forms, a gelatinous material, produced by adding the stoichiometric quantity of alkali to a cold beryllium salt solution, and a granular form obtained by making the precipitation from a hot solution, or by boiling a beryllate solution. Beryllium hydroxide can also be prepared by the electrolysis of beryllium salt solutions; it is a cathodic product, and its form depends upon the conditions[25].

From solubility measurements of beryllium hydroxide in alkali, Fricke and Humme[26] have shown that the gelatinous form is converted first into an unstable, then into a stable, crystalline form; the solubility in alkali decreases as these changes in crystalline form occur. Other work of Fricke[27, 28] supports his contention that there are two crystalline forms of the hydroxide. X-ray examination of the stable crystalline form proves it to have an orthorhombic cell, similar to Σ $Zn(OH)_2$, in which the OH groups are arranged tetrahedraly around the Be^{2+} cation[29]. In this respect beryllium hydroxide differs from most other crystalline hydroxides; these all possess layer lattices. The difference may arise from the high field strength of the Be^{2+} ion which draws the hydroxyl groups into a rather more close-packed structure than a layer lattice allows, and also from the inability of beryllium to exercise a coordination number greater than four.

Infrared absorption spectra of the stable crystalline form of beryllium hydroxide have been interpreted as showing that hydrogen bonding occurs as in other crystalline hydroxides[30]. A plot of the absorption maxima of fourteen crystalline hydroxides against the hydroxyl radius deduced from crystal measurements is a smooth curve except for the maxima of Σ $Zn(OH)_2$ and $Be(OH)_2$. The exceptional behaviour of these two hydroxides is attributed to the abnormal polarization of their hydroxyl groups. This observation provides support for the explanation of the abnormal crystal structure of Σ $Zn(OH)_2$ and $Be(OH)_2$, given in the previous paragraph.

When heated, beryllium hydroxide loses water, but quite a high temperature is required to complete its dehydration to oxide. Thus Duval[31] has shown with the thermobalance that temperatures $\rangle 950°$ are required for the complete conversion of analytical samples of $Be(OH)_2$ to BeO, although lower temperatures of 500—900° may be used in larger scale operation. There is, however, evidence of beryllium hydroxide in the vapour phase at 1200° and above[32-34]; thus, for example, water vapour reacts with BeO, $BeO.Al_2O_3$ and $BeO.3Al_2O_3$ between 1300—1600° and beryllium hydroxide sublimes. The partial pressure of the latter over beryllium oxide, is 0.55 mm. at 1550°[34].

Further evidence for the existence of beryllium hydroxide at high temperatures is provided by von Wartenberg[35], who showed that molten beryllium oxide absorbs water, releasing it upon solidification. He considered that beryllium hydroxide at $\rangle 2000°$, exists both in the gaseous phase and in solution in the molten oxide. These results indicate the need for care when considering the range of conditions under which compounds can exist; these may be very different from those deduced from work under normal laboratory conditions.

References p. 21

4. *Acidic reactions of beryllium hydroxide*

Although the amphoteric properties of beryllium hydroxide have long been known, only relatively recently has the chemistry of beryllates received much attention. Fricke and Humme[26], from solubility measurements of the stable crystalline form of beryllium hydroxide in alkali, showed that the hydroxide is the equilibrium solid phase up to *c* 35% sodium hydroxide, at which point $NaBeO_2.H_2O$ becomes the equilibrium solid phase. In addition to this material a sodium beryllate Na_4BeO_3 has been obtained by direct interaction of Na_2O and BeO at 500°[36].

The presence of hydroxyl bridge bonds in the $Be_3(OH)_3^{3+}$ ion has been suggested by Kakihana and Sillén[17] (see p. 10), and Thomas and Miller[37] also suggest that similar hydroxyl bridge bonding (olation) occurs in the colloidal particles formed in solution before beryllium hydroxide is precipitated from acid solution, and in gelatinous beryllium hydroxide. The latter workers have produced stable anionic beryllium oxide hydrosols in low concentrations of strongly coordinating anions such as citrate. The slow ageing of such sols is ascribed to the loss of hydroxyl groups by aquotization thus:

$$\left[R\equiv Be-OH-\underset{\underset{OH}{|}}{\overset{\overset{OH}{|}}{Be}} -OH-\underset{\underset{OH}{|}}{\overset{\overset{OH}{|}}{Be}} -OH-\underset{\underset{OH}{|}}{\overset{\overset{OH}{|}}{Be}} -OH_2 \right]^{3-} + H_2O =$$

$$\left[R\equiv Be-OH-\underset{\underset{OH}{|}}{\overset{\overset{OH}{|}}{Be}} -OH-\underset{\underset{OH}{|}}{\overset{\overset{OH}{|}}{Be}} -OH-\underset{\underset{OH}{|}}{\overset{\overset{OH_2}{|}}{Be}} -OH_2 \right]^{2-} + OH^-$$

Similar olated anionic complexes are believed to be formed when beryllium hydroxide is dissolved in alkali[38]. Addition of alkali to the uncharged gelatinous hydroxide first displaces water molecules at the end of the olated beryllium hydroxide units to give negatively charged colloidal beryllate particles. At higher alkali concentrations, the hydroxyl bridges are progressively broken down by the reaction:

$$\text{Be}(\mu\text{-OH})_2\text{Be} + 2\text{OH}^- \rightleftharpoons \text{Be(OH)}_2 + \text{Be(OH)}_2$$

This process leads to progressively smaller polymers with increased charge/beryllium ratios, the final limit of the reaction being reached with the formation of the mononuclear beryllate anion $[\text{Be(OH)}_4]^{2-}$ (or BeO_2^{2-}). To effect a complete conversion of polynuclear beryllates to the mononuclear species requires a high alkali concentration; for example diffusion experiments[39] show that in solutions with an $\text{Na}_2\text{O/BeO}$ mole ratio as high as 14 : 1 the predominant beryllate species contains ten beryllium atoms.

The beryllate system resembles other isopolyacid systems in that an increasing concentration of alkali on the negative side of the isoelectric point decreases the degree of polymerization in the system. Many of the reactions of beryllates are dependent upon this extent of polymerization, and, accordingly, upon the $\text{Na}_2\text{O/BeO}$ mole ratio. Thus the precipitation of granular beryllium hydroxide when a beryllate solution is boiled, an operation of industrial

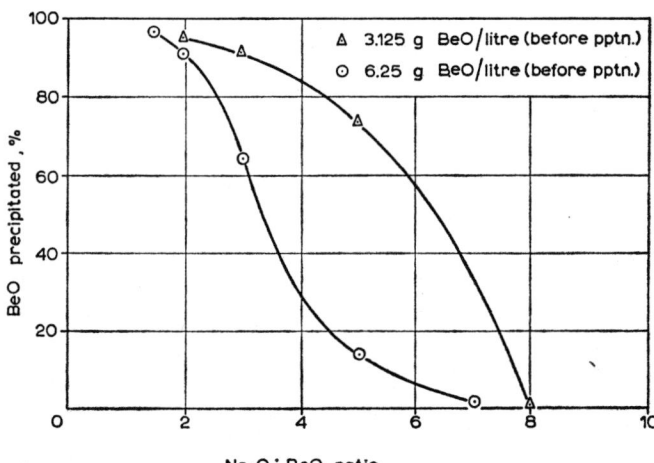

Fig. 2. Stability of beryllate solutions at 95° (after 2 h heating). Effect of the Na_2O:BeO mole ratio.
(Reproduced, with permission, from *J. Inorg. Nucl. Chem.* 24 (1962) 525.)

importance (see Chapter 8), decreases as the Na_2O/BeO mole ratio is raised (see Fig. 2). This is to be expected, since the large polymeric beryllate species present at lower alkali concentrations can more readily coalesce because of their lower charge/beryllium ratio than can mononuclear $[Be(OH)_4]^{2-}$ ions.

Addition of calcium ions to a solution of a polynuclear beryllate precipitates a complex calcium beryllate with a calcium to beryllium mole ratio of about 1 : 4—5[38]. This material has a layer lattice structure; the calcium ions are probably sandwiched between the beryllium hydroxide layers, and are the cause of the beryllium hydroxide units assuming a layer rather than a tetrahedral structure. Variations of the Ca/Be mole ratio in the compound have little effect on the unit cell size. Brandenberger[40] found the same for the $xCa(OH)_2.yAl_2O_3.zH_2O$ system in which variations in x, y and z have little influence on the X-ray diffraction patterns.

5. Interactions of beryllium cations with ion-exchange resins and solvents

As for its other chemical properties, it is the high charge to radius ratio of the Be^{2+} ion which accounts for its affinity for ion-exchange resins and solvents. Its affinity for a sulphonic acid cation-exchange resin is usually considered to be less than that of the other alkaline earth cations. However, Bonner et al.[41] find the affinity order to be $Mg^{2+}\langle Be^{2+}\langle Ca^{2+}\langle Sr^{2+}\langle Ba^{2+}$ from chloride media; this they ascribe to part of the beryllium being taken up by the resin as the unipositive species $BeCl^+$ (cf. ref. 42), or as the unipositive hydrolytic species $(BeOH)_n^{n+}$ (cf. ref. 20). In relation to the latter it is of interest that the effectiveness of the alkaline earth nitrates as salting out agents in the solvent extraction of uranyl nitrate is in the order $Mg^{2+}\rangle Be^{2+}\rangle Ca^{2+}$ etc. The anomalous position of Be^{2+} is explained[43] by the ion being extensively hydrolysed in solution; this hydrolysis leads to a fall in the charge to radius ratio and hence to a decrease in its dehydrating power.

The low affinity of the Be^{2+} ion for a sulphonic acid cation exchanger is compatible with its high degree of hydration[44], or, put in another way, with the opposite order-disorder effects of the Be^{2+} and the $R-SO_3^-$ ions. Be^{2+} is an order-producing ion, whereas $R-SO_3^-$ is a disorder-producing ion because of its low charge to radius ratio[45]. However, on phosphonic acid resins the affinities of the alkaline earth elements is reversed, Be^{2+} now being the most strongly held ion of the group. This reversal of affinities is ascribed to localised hydrolysis[45,46], whereby it is suggested[47] that an ion, such as Be^{2+}, which strongly polarizes water molecules in its vicinity, can be associated with a proton accepting anion derived from a weak acid by the mechanism $Be^{2+}\ldots \overset{\delta-}{O}H\ldots \overset{\delta+}{H}\ldots X^-$; in this the cation and anion are linked by a water bridge. As phosphonic acid groups are much weaker acids than sulphonic acid groups the high affinity of Be^{2+} for the former is readily explained. Similar reasoning can account for the high affinity of beryllium for phosphate ester ion-exchange solvents (*vide infra*).

As would be expected from this discussion, beryllium is selectively absorbed by phosphonic acid resins from solutions containing non-complex-forming anions, such as sulphate[20,48]. The commercial resin Duolite C-61 containing $RPO(OH)_2$ groups or the sodium form of a diallyl phosphate polymer[49] containing $(RO)_2PO.OH$ groups are reasonably satisfactory for this purpose. Of all the common cations, aluminium (Al^{3+} has a charge to radius ratio approaching that of the Be^{2+} ion) and ferric iron compete most strongly. However, this interference, and that of most other cations, can be suppressed by addition of E.D.T.A.[48,50], which complexes very weakly with beryllium under the mildly acidic conditions employed (pH 2—5)[51].

The solvent extraction of beryllium from acidic media by the ion-exchange type alkyl phosphoric acid extractants, dissolved in a non-polar solvent such as benzene or kerosene, has been studied on the tracer level (^{7}Be) by Hardy, Greenfield and Scargill[52]. For a dialkyl-phosphoric acid, HR, the extraction can be represented thus:

$$Be^{2+}_{aq} + 2(HR)_{2\,(org)} \rightleftharpoons BeR_2(HR)_{2\,(org)} + 2H^+_{aq}$$

References p. 21

This equation holds for low loadings of the organic phase, that is when an excess of HR is present, and indicates that the extraction of beryllium is repressed as the acid concentration of the aqueous phase is raised to above 3 M. At higher concentrations the extraction of beryllium increases probably owing to its changing to a non ion-exchange mechanism in which the phosphate ester acts similarly to the straight-complexing tributylphosphate. At the lower acidities, the extraction of beryllium is reduced by the presence of complexing anions in the solution, fluoride for instance (see p. 43).

The structure of the alkyl groups in the di-alkylphosphoric acid extractants has a slight effect on the equilibrium value

$$D_{Be} = \frac{\text{conc. Be in organic phase}}{\text{conc. Be in equal vol. aqueous phase}};$$

the value is lower the more branched the alkyl groups for a given number of carbon atoms, thus 2-ethylhexyl $\langle$ *iso*-octyl $\langle$ *n*-octyl. Also there is a similar increase in D_{Be} with chain length for *n*-alkyl groups, thus *n*-butyl $\langle$ *n*-octyl $\langle$ *n*-decyl. The rate at which equilibrium is attained with beryllium is slower than with other bivalent cations; the rate decreases with increase of the acid concentration in the aqueous phase. For individual esters the rate of attainment of equilibrium is lower the longer or more branched the alkyl chain. This has been ascribed[52] to steric hinderence to the attachment of a number of HA or A$^-$ around the small Be^{2+} cation and to the need to displace the strongly bound water molecules from the primary hydration layer of the Be^{2+} ion. However, the fact that the rate of reaction increases with increasing pH for any particular ester, probably indicates that hydrolytic beryllium species with a lower charge to radius ratio (and hence lower hydration) may also be involved in the reaction at higher pH values. Incidentally, the rate with the same alkylphosphoric acid extractants for cations of lower charge to radius ratio than Be^{2+}, for instance, UO_2^{2+}, is appreciably more rapid.

The extraction of macro concentrations of beryllium from acidic sulphate solutions by di-(2-ethylhexyl)hydrogen phosphate (HX) has been studied by Wells *et al.*[53] and by Cattrall[54], who confirm that the rate of extraction of beryllium is slow relative to that of other bivalent metal ions. The results of Wells *et al.* could be interpreted in terms of the extraction of the $BeX_2(HX)_2$ species at low solvent loadings, but with an excess of beryllium present in the aqueous phase the solvent loading approaches the theoretical limit of BeX_2. As this composition is approached the organic phase becomes viscous, suggesting the presence of long chain polymers, similar to

$$\left[\begin{array}{c} RO \diagdown \diagup OR \\ P \\ O \diagup \diagdown O \end{array} \right]$$

[chemical structure diagram of beryllium phosphate polymer with repeating unit n]

those postulated by Baes *et al.*[55] for the UO_2X_2 species. Confirmation of the polymeric character of the BeX_2 species is given by the noticable Tyndal effect exhibited by these viscous organic solutions.

The ratio HX/Be in the organic phase can drop below two when the beryllium concentration in the aqueous phase is high and the pH rises above 3. The highest ratio (~ 1.5) is obtained when the aqueous phase at equilibrium contains about 10 g BeO per 1. and has a pH of 5.1. A beryllium loading of the organic phase beyond BeX_2 is accompanied by a reduction in the light scattering and viscosity of the organic phase. These results are interpreted[53] as showing that a basic, possibly polynuclear, species is now being extracted, and that this species cannot form large polymeric structures in the organic phase as does the BeX_2 complex. It should be noted that the amount of aluminium extracted by the HX-kerosene solutions never exceeds the AlX_3 ratio, even under comparable conditions to those affording extraction of the $BeX_{1.5}$ species.

References p. 21

6. Beryllium in liquid ammonia

Owing to liquid ammonia being a more basic solvent than water, the acidic properties of beryllium are more marked in it. As shown by Bergström[56], metallic beryllium reacts with potassamide dissolved in liquid ammonia to give a blue solution of metallic potassium:

$$Be + 2KNH_2 = Be(NH_2)_2 + 2K$$

The apparent anomaly of beryllium replacing the more electropositive potassium is a mass-action consequence of the low solubility of beryllium amide, the precipitation of which forces the reaction to the right. But, it is only possible to obtain impure beryllium amide in this way as beryllium metal catalyses the reaction of the liberated potassium with the solvent:

$$2K + 2NH_3 = 2KNH_2 + H_2$$

The potassamide so formed dissolves the amphoteric beryllium amide to give potassium aminoberyllate which can be isolated by evaporating the resulting solution:

$$Be(NH_2)_2 + KNH_2 + NH_3 = K[Be(NH_2)_3NH_3]$$

This should probably be formulated as indicated, with ammonia filling the fourth coordination position of the beryllium when in solution in liquid ammonia. It is colourless and readily hydrolysed by water.

In contrast to the behaviour of beryllium halides in the water system, complex beryllium halides are obtained in the ammonia system. Beryllium dissolves in a solution of ammonium iodide in liquid ammonia to give beryllium iodide:

$$Be + 2NH_4I = BeI_2 + 2NH_3 + H_2$$

This solution dissolves more beryllium to produce compounds formulated as $3Be(NH_2)_2 \cdot BeI_2 \cdot 4NH_3$ and $5Be(NH_2)_2 \cdot BeI_2 \cdot 4NH_3$. Similar materials are also formed when an excess of beryllium is dissolved in liquid ammonia solutions of ammonium chloride or bromide, the tendency to form such complex compounds falls from iodide to chloride. Although these materials have not been fully characterised, there appears little doubt of their existence.

SOLUTION CHEMISTRY

REFERENCES

1 GURNEY, R. W., *Ionic Processes in Solution*. McGraw-Hill, New York, 1953.
2 LATIMER, W. H., *Oxidation Potentials*. 2nd. ed. Prentice Hall. New York, 1952.
3 KAMINSKY, M., *Disc. Farad. Soc.*, 24 (1957) 171.
4 FRICKE, R. AND SCHWETZDELLER, H., *Z. Anorg. Chem.*, 131 (1923) 130.
5 PASYNSKI, A., *Acta Physicochim. U.R.S.S.*, 8 (1938) 385; SPANDAU, H. AND SPANDAU, G., *Z. Phys. Chem.*, 192 (1943) 211.
6 BASOLO, F., in *Chemistry of the Coordination Compounds*, BAILAR, J. C., Ed. Reinhold, New York, 1956.
7 CARTLEDGE, G. H., *J. Amer. Chem. Soc.*, 50 (1928) 2855; idem, ibid., 52 (1930) 3076.
8 SCHOLDER, R. AND PATSCH, R., *Z. Anorg. Chem.*, 222 (1935) 135.
9 PARSONS, C. L., ROBINSON, W. O. AND FULLER, C. T., *J. Phys. Chem.*, 11 (1907) 651.
10 SIDGWICK, N. V. AND LEWIS, N. B., *J. Chem. Soc.*, (1926) 1287.
11 PRYTZ, M., *Z. Anorg. Chem.*, 180 (1929) 355; idem, ibid., 197 (1931) 103; idem, ibid 231 (1937) 238.
12 SCHAAL, R. AND FAUCHERRE, J., *Bull. Soc. Chim. France*, (1947) 927.
13 TEYSSEDRE, M. AND SOUCHAY, P., ibid., (1951) 945.
14 SOUCHAY, P., ibid., (1948) 143.
15 MATTOCK, G., *J. Amer. Chem. Soc.*, 76 (1954) 4835.
16 GARRETT, A. B., *Helv. Chim. Acta*, 43 (1960) 2176.
17 KAKIHANA, H. AND SILLÉN, L. G., *Acta Chem. Scand.*, 10 (1956) 985; see also CARELL, B. AND OLIN, A., ibid., 15 (1961) 1875.
18 SILLÉN, L. G., *Quart. Rev.*, 13 (1959) 146.
19 BJERRUM, J., SCHWARZENBACH, G. AND SILLÉN, L. G., *Stability Constants*. Chemical Society, London, 1957.
20 AVESTON, J. AND MILWARD, G. L., *Scientific Report, N.C.L./A.E. 185*, The National Chemical Laboratory (1959); see also A. I. ZHUKOV et al., *Zhur. Neorg. Khim.*, 7 (1962) 1448.
21 SIDGWICK, N. V., *Chemical Elements and Their Compounds*. Vol. 1 (p. 211), Oxford University Press, 1950.
22 DIEBLER, H. AND EIGEN, M., *Z. Phys. Chem. Neue Folge*, 20 (1959) 299.
23 CZERLINSKI, G. AND EIGEN, M., *Z. Electrochem.*, 63 (1959) 652; EIGEN, M. *Disc. Faraday Soc.*, 17 (1954) 194.
24 EIGEN, M., *Proc. Sixth International Conf. Coord. Chem.*, Detroit, (1961) p. 371, Macmillan, New York, 1961.
25 PARIKH, R. K. AND KAMMERMEYER, K., *Ind. Eng. Chem.*, 45 (1953) 1585; DOMINE-BERGER, M., *Ann. Chim. France* (1950) 106.
26 FRICKE, R. AND HUMME, H., *Z. Anorg. Chem.*, 178 (1929) 400.
27 HAVESTAT, L. AND FRICKE, R., ibid., 188 (1930) 357.
28 FRICKE, R. AND SEVERN, H., ibid., 205 (1932) 287.
29 SEITZ, A., ROSLER, U. AND SCHUBERT, K., ibid., 261 (1950) 94.
30 HARTERT, E. AND GLEMSER, O., *Naturwiss.*, 40 (1953) 199; GLEMSER, O. AND HARTERT, E., *Z. Anorg. Chem.*, 283 (1956) 111.

31 DUVAL, C., *Anal. Chim. Acta*, 1 (1947) 53.
32 HUTCHISON, C. A. AND MALM, J., *J. Amer. Chem. Soc.*, 71 (1949) 1338.
33 GROSSWEIMER, L. I. AND SIEFERT, R. L., *ibid.*, 74 (1952) 2701.
34 YOUNG, W. A., *J. Phys. Chem.*, 64 (1960) 1003.
35 VON WARTENBERG, H., *Z. Anorg. Chem.*, 264 (1951) 226; idem, *Z. Electrochem.*, 55 (1951) 445.
36 ZINTL, E. AND MORAWIETZ, A., *Z. Anorg. Chem.*, 236 (1938) 372.
37 THOMAS, A. W. AND MILLER, H. S., *J. Amer. Chem. Soc.*, 58 (1936) 2526.
38 EVEREST, D. A., MERCER, R. A., MILLER, R. P. AND MILWARD, G. L., *J. Inorg. Nucl. Chem.*, 24 (1962) 525.
39 BRINTZINGER, E. AND OSSWALD, H., *Z. Angew. Chem.*, 47 (1934) 61.
40 BRANDENBERGER, E., *Schweiz., Min. Petr. Mitt.*, 13 (1933) 569.
41 BONNER, O. D., JUMPER, C. F. AND ROGERS, O. C., *J. Phys. Chem.*, 62 (1958) 6.
42 OHTAKI, H. AND YAMASAKI, K., *Bull. Chem. Soc. Japan*, 31 (1958) 6.
43 YAKIMOV, M. A. AND NOSOVA, N. F., *Zhur. Neorg. Khim.*, 6 (1961) 208.
44 KRESSMAN, T. E. AND KITCHENER, J. A., *J. Chem. Soc.*, (1949) 208.
45 EVEREST, D. A., *Scientific Report NCL/AE 201*, The National Chemical Laboratory (1962).
46 KENNEDY, J. AND WHEELER, V. J., *Chem. and Ind.*, (1959) 1577.
47 ROBINSON, R. A. AND HARNED, H. S., *Chem. Rev.*, 28 (1941) 49.
48 KENNEDY, J. AND WHEELER, V. J., *Anal. Chim. Acta*, 20 (1959) 412.
49 KENNEDY, J., LANE, E. S. AND ROBINSON, B. I. C., *J. Applied Chem.*, 8 (1958) 459.
50 NADKARNI, M. N., VARDE, M. S. AND ATHERVALE, V. T., *Anal. Chim. Acta*, 16 (1957) 421.
51 FAIRHALL, A. W., *Radiochemistry of Beryllium, U.S.A.E.C. report NAS-NS 3013* (1960).
52 HARDY, C. J., GREENFIELD, B. F. AND SCARGILL, D., U.K.A.E.A. report AERE-R3316 (1960); idem, *J. Chem. Soc.*, (1961) 174.
53 WELLS, R. A., EVEREST, D. A. AND NORTH, A. A., *Nuclear Science and Engineering*, 1963, in the press.
54 CATTRALL, R. W., *Australian J. Chem.*, 14 (1961) 163.
55 BAES, C. F., ZINGARO, R. A., AND COLEMAN, C. F., *J. Phys. Chem.*, 62 (1958) 129.
56 BERGSTROM, F. W., *J. Amer. Chem. Soc.*, 50 (1928) 657.

CHAPTER 3

Simple Oxosalts of Beryllium

Here we consider the preparation, properties and reactions of a number of the simple oxosalts of beryllium which are more or less well-known. Those of the halides are sufficiently important to justify separate treatment (p. 38). Again it is the high charge to radius ratio of the Be^{2+} ion which is a root cause of the marked differences between the properties of beryllium salts and those of the other alkaline earth elements. Unfortunately, information about many beryllium oxosalts is relatively scanty, certainly as compared to other branches of beryllium chemistry, and the gaps make it difficult to present a coherent picture.

A practical difficulty in the way of their study arises from the very high solubility in water of beryllium oxosalts, which makes their preparation difficult. The difficulty is increased by the formation of hydrolytic beryllium species; these usually make it desirable to carry out preparative work in solutions on the acid side of neutrality. It must be emphasised that the hydrolytic polymerisation reactions of the Be^{2+} ion described previously occur readily in solutions of all beryllium salts; these will only be mentioned here when they bear on other aspects of beryllium chemistry under discussion.

1. Beryllium carbonates

These compounds are considered first not because of their intrinsic importance but because beryllium oxide carbonate is commonly used as an intermediate in the preparation of other salts. The oxide carbonate is precipitated when sodium carbonate is added to a beryllium salt solution, carbon dioxide being evolved owing to the acidity of such solutions resulting from the hydrolysis of the Be^{2+} ion. The precipitate is considered to be a mixture of the normal carbonate, $BeCO_3$, and beryllium hydroxide[1-3]; it usually contains from two to five molecules of $Be(OH)_2$ to every one of $BeCO_3$, the ratio of hydroxide to carbonate increasing as the precipitation temperature is raised. On heating, the freshly precipitated material begins to lose water at 80° and carbon dioxide above 300° (values of 100–140° and 190–260° have also been given for these decomposition temperatures)[4]. This separate evolution of water and carbon dioxide is consistent with the precipitate being a mixture rather than an individual compound.

Beryllium carbonate, $BeCO_3.4H_2O$, has been prepared by passing carbon dioxide through an aqueous suspension of beryllium hydroxide[4,5]; it is unstable and obtained only when the solution is under pressure of carbon dioxide. X-ray analysis shows $BeCO_3.4H_2O$ to be hexagonal, a = 5.12, b = 15.77Å. As with other beryllium salts, the order of thermal decomposition of the alkaline earth carbonates is Be⟩⟩Mg⟩Ca⟩Sr⟩Ba (see p. 25 for an explanation of this effect).

Beryllium hydroxide dissolves in sodium bicarbonate solution[6-8], differing in this from aluminium hydroxide which is insoluble; this has been considered as a means for the large scale separation of beryllium from aluminium[6]. Little is known about the beryllium species formed except that they are probably anionic, since beryllium is taken up by anion-exchange resins from these solutions[8]. Addition of alcohol results in a precipitate with the approximate composition $2NaHCO_3.Be(OH)_2$[7]. There is not much information about this material, although the corresponding aluminium and gallium compounds appear well established[9]. These are prepared by interaction of sodium bicarbonate with sodium aluminate or gallate, formulated as $Na_2O.R_2O_3.2CO_2.nH_2O$, and shown to be individual compounds by X-ray and thermal analysis.

The soluble complex beryllium carbonate can also be obtained by dissolving beryllium oxide carbonate or hydroxide in ammonium carbonate solution[6,10]. Dissolution is rather slow and a large excess of ammonium carbonate is required for complete dissolution of the beryllium.

2. *Beryllium sulphate*

This is the most important and the most completely studied oxosalt of beryllium. It is most frequently met as the tetrahydrate, $BeSO_4.4H_2O$, prepared by evaporating a solution of beryllium oxide carbonate or hydroxide in dilute sulphuric acid. A number of other hydrates have been described but the di-, tetra- and, possibly, the pentahydrate are definitely established. Campbell, Sukava and Coop[11] state that the transitions on dehydration are $BeSO_4.4H_2O \xrightarrow{89°} BeSO_4.2H_2O \xrightarrow{270°} BeSO_4$, but Kraus and Gerlach[12] from tensimetric measurements, consider them to occur at 120° and 400° respectively. Novoselova and Levina[13] confirm that $BeSO_4.2H_2O$ is the only intermediate hydrate, its transition to the anhydrous form being said to occur at 250°. Claims have been made for tri-, mono- and hemi-hydrates in the $BeSO_4-H_2O$ system, from thermal analysis with a slow rate of heating (5°/min.)[14], but these results lack confirmation. Evidence has been obtained for the formation of a pentahydrate[15] below $-16°$; in it the extra water molecule may be associated with the sulphate anion as in $CuSO_4.5H_2O$.

Anhydrous beryllium sulphate can be readily prepared by dehydrating the tetrahydrate in a vacuum at about 250°, or by the direct action of dimethyl sulphate at 160–190° on beryllium oxide in absence of moisture[16]. On heating it begins to decompose at 580-600°, one hour at this temperature causing about 4% of the sulphur to be evolved as SO_3[17], the vapour pressure of sulphur trioxide over beryllium sulphate being 365 mm at 750°[18]. As the decomposition of aluminium sulphate is appreciably more rapid (vapour pressure of sulphur trioxide over aluminium sulphate is 900 mm at 750°) it is possible to separate aluminium and beryllium by differential desulphatization.

Although beryllium sulphate has a moderate thermal stability, it is appreciably less stable than the other alkaline earth sulphates, magnesium sulphate for instance is stable up to a white heat. The difference is because the strongly polarising Be^{2+} ion deforms the sulphate anion and thereby loosens the sulphur–oxygen bonds; this in turn leads to a lower thermal stability of the sulphate radicle (cf. ref. 19). In this connection it is significant that Heerdt and Gorbeau[20], showed, by Raman spectra measurements on crystalline sulphates, that the sulphur-oxygen vibration frequencies increased with the charge to radius ratio of the cation. Similar conclusions concerning the deformation of the sulphate anion were reached by Duval and Lecompte[21] from infrared measurements.

The structure of beryllium sulphate tetrahydrate has been shown by X-rays[22, 23] to consist of tetrahedral $[Be(H_2O)_4]^{2+}$ cations and sulphate anions arranged in approximately a body centred, cubic, caesium chloride structure. Every water molecule has two external contacts to the oxygen atoms of the sulphate anion. Anhydrous beryllium sulphate is reported to possess a cristobalite structure[24].

Beryllium sulphate tetrahydrate is readily soluble in water, its solubility being 42.5 and 37 g per 100 g water at 25° and 18° respectively[25], its solubility increasing markedly in presence of beryllium oxide[26-28]. Its solubility is decreased in presence of sulphuric acid[11]. At high acid concentrations the tetrahydrate is not the stable phase even at 25°, thus X-rays show the solid phase in equilibrium with 80–98% (by weight) sulphuric acid to be a mixture of the dihydrate and the anhydrous salt[29]. Beryllium sulphate has a negligible solubility in 100% sulphuric acid and it does not exceed 2.5% $BeSO_4$ in the range 80–98% (by weight) sulphuric acid. In such concentrated acid measurement is made difficult by the formation of supersaturated metastable solutions of beryllium sulphate[29].

Beryllium sulphate forms double salts with sodium, potassium and ammonium sulphates. In the Na_2SO_4–$BeSO_4$–H_2O, K_2SO_4–$BeSO_4$–H_2O and $(NH_4)_2SO_4$–$BeSO_4$–H_2O systems[30] the compounds $3Na_2SO_4.BeSO_4$, $K_2SO_4.BeSO_4.2H_2O$ (see also ref. 31), $(NH_4)_2SO_4.BeSO_4.2H_2O$ (below 50°) and $(NH_4)_2SO_4.BeSO_4$ (above 50°, see also ref. 32) have been isolated. Sidgwick[33] has

$$M_2\left[\begin{matrix} O & O & O & O \\ \diagdown S \diagdown & \diagup Be \diagdown & S \diagup \\ O & O & O & O \end{matrix}\right]$$

suggested a dichelate structure for these compounds, the two water molecules of crystallisation being associated with the sulphate anions. However, X-ray evidence for or against the occurence of such structures is at present lacking.

Beyond the elements mentioned above double salts with beryllium sulphate are not found; in particular not in the systems Li_2SO_4–$BeSO_4$–H_2O[34], Ag_2SO_4–$BeSO_4$–H_2O[35], $FeSO_4$–$BeSO_4$–H_2O[36], $CuSO_4$–$BeSO_4$–H_2O[37], $CaSO_4$–$BeSO_4$–H_2O[38] and $Al_2(SO_4)_3$–$BeSO_4$–H_2O[39]. Study of the quaternary system $(NH_4)_2SO_4$–$BeSO_4$–$Al_2(SO_4)_3$–H_2O[39] revealed that $(NH_4)_2SO_4$.$BeSO_4$.$2H_2O$ and ammonium alum were the only binary salts formed at room temperature. This system is of significance in the separation of beryllium and aluminium on an industrial scale (p. 111).

The only anhydrous beryllium sulphate system so far studied is that of Na_2SO_4–$BeSO_4$[40]. Thermal analysis measurements showed the formation of the three binary compounds Na_2SO_4.$BeSO_4$, $3Na_2SO_4$.$BeSO_4$ and Na_2SO_4.$3BeSO_4$. The compound $3Na_2SO_4$..$BeSO_4$ has also been reported in aqueous systems[30].

3. Beryllium selenate

Beryllium selenate tetrahydrate, $BeSeO_4$.$4H_2O$, is prepared by evaporating a solution of beryllium oxide carbonate in a slight excess of selenic acid[26]. It is readily soluble in water and resembles beryllium sulphate in its ability to dissolve beryllium oxide[26, 41]. Differential thermal analysis of the tetrahydrate[42] shows strong endothermic effects at 75°, 146°, 213° and 560–610°. That at 75° is ascribed to the transition $BeSeO_4$.$4H_2O \to BeSeO_4$.$2H_2O$, those at 146° and 213° to the transition $BeSeO_4$.$2H_2O \to BeSeO_4$, and that at 560-610° to a breakdown of the salt to beryllium oxide and the decomposition products of selenium trioxide.

Although the dehydration of $BeSeO_4$.$4H_2O$ tends to cause volatilisation and partial reduction of the selenium, the dihydrate has been obtained in reasonable purity by standing the tetrahydrate in a desiccator over phosphorous pentoxide, and the anhydrous salt by heating the dihydrate in a vacuum at 215–220°[43]. As with the sulphate[44], the heat of combination of beryllium selenate with the first pair of water molecules is much greater than with the second pair[43]. The lower stability of the selenate relative to the sulphate is reflected in the lower heat of formation of the former (-213.26 as against -280.6 kcal. mole^{-1}).

4. Beryllium nitrate

The tetrahydrate, $Be(NO_3)_2.4H_2O$, can be prepared by crystallizing a solution of beryllium hydroxide or carbonate from a slight excess of dilute nitric acid[45,46]. The crystallisation of beryllium nitrate from more concentrated nitric acid solutions leads to formation of tri-, di-, and mono-hydrates. Tarem[47] claims that the anhydrous salt is obtained by heating the trihydrate to 160°, decomposition of the salt occuring at 170°. But this has been contested by Novoselova[48] who states that the trihydrate decomposes at 160° with evolution of oxides of nitrogen. Indeed, it appears doubtful whether the anhydrous salt can be obtained by dehydration of hydrated beryllium nitrate; for example attempts to dehydrate the tetrahydrate at 56° over solid potassium hydroxide result in loss of N_2O_5.

Differential thermal analysis of the tetrahydrate[49] shows that the salt first melts in its own water of crystallization and then decomposes with the evolution of nitrous fumes; the second reaction begins at relatively low temperatures and appears complete by 250°. These results support Novoselova's observations rather than those of Tarem. As with the other hydrated alkaline earth nitrates, the thermal decomposition of $Be(NO_3)_2.4H_2O$ occurs in such a way that nitric acid is lost before nitrous fumes are evolved, that is hydrolysis preceeds thermal decomposition.

Anhydrous beryllium nitrate has been obtained by solvolysis of beryllium chloride in ethylacetate-dinitrogen tetroxide mixtures to give pale-straw coloured crystals of $Be(NO_3)_2.2N_2O_4$[50]. When heated in vacuum this compound decomposes in two stages. Dinitrogen tetroxide is evolved rapidly above 50° leaving anhydrous beryllium nitrate. The latter compound is stable to *ca.* 125° when a sudden decomposition occurs to give dinitrogen tetroxide and a volatile beryllium compound which separates from the gas phase as colourless crystals of composition $Be_4O(NO_3)_6$, analagous to beryllium oxide acetate (see p. 71).

Beryllium nitrate is very soluble in water, the solution showing the usual hydrolytic reactions of the Be^{2+} ion. Claims have been made for the detection of autocomplex formation in beryllium nitrate solutions from interfacial measurements[51], but such interactions must be small as, in contrast to many cations, Be^{2+} has little effect on the Raman spectrum of the aqueous nitrate ion[52].

5. Beryllium phosphates

Despite the general importance of the beryllium phosphates, especially in the analytical chemistry of beryllium, there are many large gaps in our knowledge of this group of compounds. In particular, a systematic study of the $BeO–P_2O_5–H_2O$ system appears to be necessary in order to define the simple beryllium phosphates and establish their stability limits. This information is of particular importance since, owing to the ready hydrolysis of the Be^{2+} ion, it is only too easy experimentally to obtain ill defined basic phosphates.

Although the mono-[53], di-[53, 54, 55, 56, 56a] and tertiary[53, 56a, 57] beryllium phosphates, $Be(H_2PO_4)_2$, $BeHPO_4$ and $Be_3(PO_4)_2$ have all been described, the best known phosphate is the compound $BeNH_4PO_4 \cdot xH_2O$ (x~1), made by adding $(NH_4)_2HPO_4$ to a beryllium solution at pH 5.2–5.5[58, 59]. When the pH is much below this some $BeHPO_4$ is also precipitated and the material contains less than the theoretical quantity of ammonia[55]. Heating $BeNH_4PO_4$ releases water of hydration at 250°, ammonia at 400° and water of constitution at 500°; at 700° conversion to $Be_2P_2O_7$ is complete[55, 60]. The loss of ammonia at 400° produces $BeHPO_4$ but the conversion is never complete and $Be_2P_2O_7$ is the only compound in this sequence[55] sufficiently specific to be analytically useful.

$BeNH_4PO_4$ crystallizes in the tetragonal system, a = b = 12.96 and c = 9.65Å, with sixteen anhydrous molecules in the unit cell.

$BeHPO_4$ is also tetragonal with a = b = 9.05 and c unchanged at 9.65Å[61].

There is evidence that phosphate forms weak complexes with beryllium in solution. Thus the formation of a $BeH_2PO_4^+$ complex is said to limit the phosphate content of solutions which can be used for the cation-exchange separation of beryllium from other elements[62]. The passage of a solution of beryllium sulphate in 90% phosphoric acid through an anion exchanger leads to both phosphate and beryllium (0.3 mM Be per g resin) being taken up by the resin, possibly because of the presence of anionic beryllium phosphate complexes[29].

6. Beryllium arsenates

These compounds are very similar to the phosphates. Beryllium ammonium arsenate, $BeNH_4AsO_4 \cdot xH_2O$, is the best known of them; it is obtained by adding ammonium arsenate to a solution of a beryllium and an ammonium salt (Cl, NO_3, SO_4) until a final pH of 6.5–7.0 is reached[63, 64]. Prepared under these conditions $x \sim 1.5$; when heated it loses water of hydration at about 250°, ammonia and water of constitution at about 400° to give a mixture of pyroarsenate $Be_2As_2O_7$ and orthoarsenate Be_3AsO_4, and finally As_2O_3 and oxygen above 900°. X-ray analysis of the material formed by heating beryllium ammonium arsenate to 1100° shows the presence of orthoarsenate and beryllium oxide, but above 1200° beryllium oxide is the only recognisable product.

Beryllium orthoarsenate is appreciably less thermally stable than the other alkaline earth arsenates; this is another illustration of the effect of the strongly polarising Be^{2+} cation on the anion with which it is combined (see p. 25).

7. Beryllium oxalate

Our knowledge of this compound is mainly drawn from the classical paper of Sidgwick and Lewis[26]. Evaporation of a solution of beryllium oxide carbonate in a slight excess of oxalic acid solution affords crystals of the trihydrate, $BeC_2O_4 \cdot 3H_2O$. This salt is peculiar in several ways. First, it is unique amongst bipositive metal oxalates in being soluble in water (24.85% of the anhydrous salt at 25°). Secondly, because it forms a trihydrate whereas beryllium salts are usually tetrahydrates. Thirdly, the behaviour of its solutions is anomalous; the conductivity of its aqueous solution is only about one quarter of that of an equivalent beryllium sulphate solution and remains at a constant value over a wide concentration range; freezing point measurements confirm that beryllium oxalate gives only a small number of ions in solution. These results were interpreted by Sidgwick and Lewis in terms of an equilibrium between an unionised hydrated monochelate complex and a dichelate complex salt:

$$2 \begin{array}{c} H_2O \\ \\ H_2O \end{array} Be \begin{array}{c} O-C=O \\ | \\ O-C=O \end{array} \rightleftharpoons Be\left[Be(C_2O_4)_2\right]$$

As metal oxalates usually have one water molecule associated with each oxalate group, either of the above species can account for the observed trihydrate. Support for the above equilibrium is provided by the fact that beryllium can be extracted by long chain amines as the anionic oxalate complex $[Be(C_2O_4)_2]^{2-}$ from these solutions[64A] and by the existence of a series of double oxalates of the type $Na_2[Be(C_2O_4)_2]$[65].

The thermal decomposition of $BeC_2O_4 \cdot 3H_2O$ has been followed by Hammer and Harris[66] as part of a study of the production of high purity beryllium oxide. They find the trihydrate to be relatively unstable in air and to decompose to monohydrate at temperatures as low as 50°. The actual course of dehydration depends upon the rate of heating employed. When the trihydrate is rapidly heated, an intermediate liquid phase is produced, but liquifaction is avoided by slow heating. Above 225° the monohydrate decomposes to beryllium oxide, the rate of decomposition increases with temperature.

References p. 35

8. *Normal beryllium salts of carboxylic acids*

Here we consider the formation and reactions of the normal beryllium salts of carboxylic acids, the beryllium oxide derivatives are dealt with on p. 71. Normal beryllium carboxylates are usually obtained under controlled conditions; particular care must be taken in drying them as only one quarter of a mole equivalent of water is required for their complete hydrolysis to the oxide salt:

$$4Be(R.COO)_2 + H_2O = Be_4O(R.COO)_6 + 2R.COOH$$

Field[67] has described the preparation of the normal formate (decomposes 150°), acetate (decomposes 294–6°), propionate (decomposes 75–78°), benzoate (decomposes 307–9°) and *o*-chlorobenzoate (decomposes 245-9°) by the action of anhydrous beryllium chloride with the acid chloride in absence of moisture. The presence of even traces of moisture lead to oxide salts. The normal acetate has also been made by treating the oxide acetate with glacial acetic acid and acetyl chloride[68]:

$$Be_4O(CH_3.COO)_6 + 2CH_3COOH + 2CH_3.COCl =$$
$$4Be(CH_3.COO)_2 + (CH_3.CO)_2O + 2HCl.$$

It is decomposed on heating:

$$4Be(CH_3.COO)_2 = Be_4O(CH_3.COO)_6 + (CH_3.CO)_2O,$$
$$\text{and } Be(CH_3.COO)_2 = BeO + (CH_3.CO)_2O.$$

This is in contrast to the decomposition of calcium acetate which gives acetone and calcium carbonate. The preparation of other normal organic acid salts of beryllium has been described[69].

The only normal beryllium salt of an aliphatic acid which is more stable than the oxide salt is the formate; in addition to Field's method, this is obtained by action of formic acid on beryllium oxide acetate[70] or by dissolving beryllium hydroxide in $\rangle 50\%$ formic acid[71]. It is an anhydrous solid, insoluble in ordinary organic solvents and only slowly hydrolysed by water. Sidgwick[33] suggests it is polymeric.

9. Beryllium perchlorate

Sidgwick and Lewis[26] described the tetrahydrate as being formed by dissolving beryllium oxide in a slight excess of perchloric acid. It is very soluble in water (59.5% at 25°) and is thermally stable, not losing water below the temperature at which the perchlorate ion begins to decompose.

10. Beryllium azide

The anhydrous salt, $Be(N_3)_2$, has been prepared by addition of ethereal hydrazoic acid to dimethylberyllium at liquid nitrogen temperatures[72]. Reaction starts at the melting point of ether (–116°), methane is evolved and $Be(N_3)_2$ precipitated as a white solid. This is obtained in a pure state after the ether and excess of hydrazoic acid have been evaporated off under vacuum. Beryllium azide is reasonably stable, detonating only slightly in a flame and being insensitive to percussion. In this respect it is similar to the other alkaline earth azides. It is soluble in tetrahydrofuran, insoluble in ether and hydrolysed by water.

11. Beryllium iodate and periodate

Hydrates of these two salts, $Be(IO_3)_2 \cdot 4H_2O$ and $Be(IO_4)_2 \cdot 8H_2O$, have been crystallized from solutions of beryllium oxide in a slight excess of the acid[73]. Like most oxosalts of beryllium they are very soluble in water. The preparation of $B_3(IO_5)_2 \cdot 12H_2O$ has been recently described.[73a]

12. Beryllium sulphonates

A number of beryllium sulphonates have been isolated from solutions obtained by dissolving beryllium oxide carbonate or hydro-

References p. 35

xide in a slight excess of the aqueous acid[26], or by double decomposition of solutions of beryllium sulphate and the barium salt of the desired acid[74]. The salts are all very soluble in water, in alcohol, acetone and hot glacial acetic acid, they are all insoluble in benzene. They generally crystallize with the usual four water molecules per beryllium atom, except p-xylenesulphonate which is said to crystallize with five water molecules[75]. When heated to 190° they are all converted to the anhydrous form; at higher temperatures they decompose e.g. $Be(C_6H_5.SO_3)_2$ at 353°, and $Be(m-C_6H_4(NO_2)SO_3)_2$ at 232°.

13. Beryllium silicates

The two most important beryllium silicates are beryl, $3BeO.Al_2O_3.6SiO_2$, and beryllium orthosilicate, phenacite, Be_2SiO_4. The preparation of these two compounds will be briefly described; further details concerning the structure and properties of beryl are given in Chapter 8.

Beryl has been synthesised hydrothermally by heating the stoichiometric quantities of alumina, silica and beryllium oxide carbonate, with sufficient water to give a pressure of about 1500 bars in a bomb at 600°. Under these conditions transparent crystals of beryl are obtained[76].

Phenacite is prepared by heating silica and beryllia to 1500° in presence of willemite, Zn_2SiO_4[77], zinc oxide, manganese dioxide or lithium carbonate[78] as the mineralising agent. When a mixture of silica and beryllia is heated to 1200–1300° in presence of an alkali fluoroberyllate[79], volatile beryllium and silicon fluorides are formed and phenacite is slowly deposited from the vapour phase on a cold surface. Phenacite has been stated in the literature to be insoluble in acids, but it has been recently found that the mineral is dissolved by boiling with 85% (by weight) sulphuric acid. Varying the acid concentration above or below this figure greatly reduces the extent of attack[80].

14. Beryllium cyanide

This compound is precipitated when on ethereal solution of of dimethylberyllium is added to an excess of hydrogen cyanide in an inert solvent, such as benzene; methane is also evolved [81]. It is insoluble in solvents other than those that cause hydrolysis, and probably has a cross-linked polymeric constitution.

REFERENCES

1 VENTURELLO, G., *Gazz. Chim. Ital.*, 69 (1939) 73.
2 TAREM, H. N., *Compt. Rend.*, 222 (1946) 1436.
3 TAREM, H. N., *Rev. Fac. Sci. Univ. Istanbul*, 11 (1946) 107.
4 SHARGORODSKI, S. D. AND SHOR, O. I., *Ukrain Khim. Zhur.*, 20 (1954) 357; *Chem. Abs.*, 49 (1955) 12932.
5 KLATZO, G., *J. Prakt. Chem.*, 106 (1869) 227.
6 LUNDIN, H., *Trans. Amer. Inst. Chem. Eng.*, 41 (1945) 671.
7 BALANDIN, A., *Chem. News*, 132 (1926) 213.
8 MERCER, R. A., MILLER, R. P. AND MILWARD, G. L., *Scientific Report NCL/AE 192*. The National Chemical Laboratory (1959).
9 PERMYAKOVA, T. V. AND LILEEV, I. S., *Zhur. Neorg. Khim.*, 5 (1960) 91 and 999.
10 MISUMI, S. AND TAKETATSU, T., *Bull. Chem. Soc. Japan.* 32 (1959) 593.
11 CAMPBELL, A. N., SUKAVA, A. J. AND KOOP, J., *J. Amer. Chem. Soc.*, 73 (1951) 2831.
12 KRAUSS, F. AND GERLACH, H., *Z. Anorg. Chem.*, 140 (1924) 61.
13 NOVOSELOVA, A. V. AND LEVINA, M. E., *Zhur. Obsch. Khim.*, 8 (1938) 1143.
14 FIALKOV, YA. A. AND SHARGORODSKI, S. D., *ibid.*, 18 (1948) 1747.
15 ROHMER, R., *Bull. Soc. Chim. France*, (1943) 468.
16 LAUTIE, R., *ibid.*, (1947) 508.
17 SUCHKOV, A. B., BOROK, B. A. AND MOROZOVA, Z. I., *Zhur. Priklad. Khim.*, 32 (1959) 1618.
18 SESHADRI, M. R. AND MALLIKARJUNAM, R. *J. Ind. Inst. Sci.*, B41 (1959) 7.
19 SPITSYN, V. I. AND SHOSTAK, V. I., *Zhur. Obsch. Khim.*, 19 (1949) 1801.
20 HEERDT, R. AND GOUBEAU, J., *Z. Anorg. Chem.*, 255 (1948) 309.
21 DUVAL, C. AND LECOMPTE, J., *Compt. Rend.*, 227 (1948) 1153.
22 BEEVERS, C. A. AND LIPSON, H., *Z. Krist.*, 82 (1932) 297.
23 WELLS, A. F., *Structural Inorganic Chemistry*, 3rd ed., Oxford University Press 1962.
24 GRUND, A., *Tschernaks Mineralog. Petrog. Mitt.*, 5 (1955) 227.
25 SCHREINER, L. AND SIEVERTS, A., *Z. Anorg. Chem.*, 224 (1935) 167.
26 SIDGWICK, N. V. AND LEWIS, N. B., *J. Chem. Soc.*, (1926) 1287.
27 PUCHE, F. AND JOSIEN, M. L., *Bull. Soc. Chim. France* (1940) 755.

28 JOSIEN, M. L., *ibid.*, (1940) 955.
29 MERCER, R. A., MILLER, R. P. AND MILWARD, G. L., in *Scientific Report NCL/AE 194*, The National Chemical Laboratory (1960).
30 SCHRODER, W. et al., *Z. Anorg. Chem.*, 239 (1938) 44 and 235; 240 (1938) 50; 241 (1939) 179.
31 BRITTON, H. T. S. AND ALLMAND, A. J., *J. Chem. Soc.*, 119 (1921) 1463.
32 BRITTON, H. T. S., *ibid.*, 121 (1922) 2612.
33 SIDGWICK, N. V., *The Chemical Elements and their Compounds*. Oxford University Press, 1950, p. 217.
34 BORIC, I. I., VOROB'EVA, O. I. AND NOVOSELOVA, A. V., *Zhur. Neorg. Khim.*, 5 (1960) 1157 and 1176.
35 ORBICH, C. A. AND EDWARDS, L. S., *J. Amer. Chem. Soc.*, 73 (1951) 4325.
36 NOVOSELOVA, A. V., VOROB'EVA, O. I., KNYAZEVA, N. N. AND PASKUTSKAYA, L. M., *Zhur. Obsch. Khim.*, 23 (1953) 1284.
37 NOVOSELOVA, A. V., BODALEVA, N. V. AND GERSHTEIN, M. M., *ibid.*, 8 (1938) 732.
38 RESKETNIKOVA, L. P., NOVOSELOVA, A. V. AND KIRKINA, D. F., *Zhur. Neorg. Khim.*, 3 (1958) 378.
39 URAZOV, G. G., KINDYAKOV, P. S. AND KHOKHLEVA, A. V., *Trudy Moskov Inst. Tonkoi Khim. Tekhnol. im M. V. Lomonosova* (1958) 90; *Chem. Abs.*, 54 (1960) 23690.
40 FEDOROV, P. I. AND CHEN-YUN CHANG., *Hua Hsueh Hsush Pao*, 23 (1957) 9; *Chem. Abs.*, 52 (1958) 13393.
41 PUCHE, F., *Bull. Soc. Chim. France*, (1950) 763.
42 SELIVANOVA, N. M., SHNEIDER, V. A. AND STREL'TSOV, I. S., *Zhur. Neorg. Khim.*, 5 (1960) 2272.
43 SELIVANOVA, N. M. AND SHNEIDER, V. A., *Zhur. Fiz. Khim.*, 35 (1961) 574.
44 ROSSINI et al., *U.S. Nat. Bur. Standards, Washington, Circ.*, 500 (1952).
45 NOVOSELOVA, A. V. AND NAGORSKAYA, N. D., *Bull. Soc. Chim. France*, (1935) 967.
46 NOVOSELOVA, A. V., NAGORSKAYA, N. D. AND METELOVA, N. M., *Zhur. Obsch. Khim.*, 6 (1936) 1306.
47 TAREM, H. N., *Compt. Rend.*, 222 (1946) 1387.
48 NOVOSELOVA, A. V., *Zhur. Obsch. Khim.*, 21 (1951) 412.
49 GORDEN, S. AND CAMPBELL, C., *Anal. Chem.*, 27 (1955) 1102.
50 ADDISON, C. C. AND WALKER, A., *Proc. Chem. Soc.*, (1961) 242; *J. Chem. Soc.*, (1963) 1220.
51 KAZE, H., *J. Ind. Chem. Soc.*, 33 (1956) 513.
52 MATHIEU, J-P. AND LOUNSBURY, M., *Disc. Farad. Soc.*, 9 (1950) 196.
53 TRAVERS, I. M. AND PERON, MLLE., *Ann. Chim. France* (1924) 298.
54 CHROMSE, H., *Z. Anorg. Chem.*, 233 (1937) 144.
55 GORIA, C. AND AIROLDI, R., *Ann. Chim. Rome*, 42 (1952) 160.
56 BOULLE, A. AND DUPIN, A. S., *Compte rend.*, 253 (1961) 2985.
56a SILBER, P. AND JAULMES, S., *ibid.*, 254 (1962) 4034.
57 BLEYER, B. AND MULLER, BR., *Z Anorg. Chem.*, 79 (1912) 263.
58 TORIBARA, T. Y. AND CHEN, P. S., *Anal. Chem.*, 24 (1952) 539.
59 HURE, J., KREMER, M AND LE BERQUER, F., *Anal. Chim. Acta*, 7 (1952) 37.

REFERENCES

60 BOUILLE, A. AND DUPIN, A. S., *Compt. Rend.*, 254 (1962) 122.
61 GORIA, C., *Atti. Acad. sci. Torino*, 92 (1957-8) 96; *Chem. Abs.*, 54 (1960) 9426.
62 FAIRHALL, A. W., *The Radiochemistry of Beryllium. USAEC report NAS-NS 3013* (1960).
63 AIROLDI, R., *Ann. Chim. Rome*, 43 (1953) 15.
64 GORIA, C. AND AIROLDI, R., *ibid.*, 43 (1953) 125.
64a DE BRUIN, H. J,, KAIRAITES, D. AND TEMPLE, R. B., *Austr. J. Chem.*, 15 (1962) 457.
65 ROSENHEIM, A. AND WAGE, P., *Z. Anorg. Chem.*, 15 (1897) 292.
66 HAMMER, R. L. AND HARRIS, L. A., *USAEC report ORNL-3183 1961*.
67 FIELD, G. W., *J. Amer. Chem. Soc.*, 61 (1939) 1817.
68 BESSON, J. AND HARDT, H. D., *Compt. Rend.*, 237 (1953) 1525.
69 PRASAD, S. AND SRIVASTAVA, K. P., *J. Ind. Chem. Soc.*, 35 (1958) 261.
70 BESSON, J. AND HARDT, H. D., *Compt. Rend.*, 238 (1954) 355.
71 TANATAR, S., *Ber.*, 43 (1910) 1230.
72 WIBERG, E. AND MICHAUD, H., *Z. Naturforsch*, 9b (1954) 502.
73 BIBER, V. A., NEUMAN, I. A. AND BROGINA, A. A., *Zhur. Obsch. Khim.*, 11 (1941) 861.
73a DRATORSKY, M. AND PREJZKOVA, J., *Coll. Czech. Chem. Comm.*, 28 (1963) 1280.
74 BOOTH, H. S. AND PIERCE, D. G., *J. Phys. Chem.*, 37 (1933) 59.
75 CUPR, V. AND SIRUCEK, J., *J. Prakt. Chem.*, 136 (1933) 159; 139 (1934) 245.
76 WYART, J. AND SCAVNICAR, S., *Bull. Soc. Franc. Miner. Crist.*, 70 (1957) 395.
77 MORGAN, R. A. AND HUMMEL, F., *J. Amer. Ceram. Soc.*, 32 (1949) 250.
78 ISUPOVA, E. N. AND KELER, E. K., *Zhur. Neorg. Khim.*, 5 (1960) 1126.
79 SOBOLOV, B. P. AND NOVOSELOVA, A. V., *Geokhimiya*, 4 (1959) 20; SOBOLOV, B. P. AND KLYAGINA, I. P., *Zhur. Neorg. Khim.*, 5 (1960) 2294.
80 *Unpublished observations*, The National Chemical Laboratory.
81 COATES, G. E. AND MUKERJEE, R. N., *J. Chem. Soc.*, (1963) 229.

CHAPTER 4

The Beryllium Halides

The most obvious feature of the chemistry of the beryllium halides is the contrast between the fluoride on the one hand and the chloride, bromide and iodide on the other. For example, beryllium forms stable fluoride complexes, but its complexes with the rest are very weak. Again, although the chloride, bromide and iodide form numerous and stable complexes with neutral ligands, the fluoride shows no tendency to do this.

The basic cause of these differences is the differing degree of ionic character of the beryllium–halogen bonds. From Paulings relationship[1] between the electronegativities of two atoms A and B and the nature of the A–B bond, it has been calculated that the Be–F bond is ionic to the extent of 80%. This is much greater than the values of 42%, 35% and 25% ascribed to the Be–Cl, Be–Br and Be–I bonds respectively. This makes beryllium–fluorine compounds essentially ionic and the other three halide compounds largely covalent.

1. Preparation of beryllium fluoride

Like all the beryllium halides, beryllium fluoride cannot be readily prepared by wet reactions, *e.g.* the dissolution of beryllium hydroxide in aqueous hydrofluoric acid, because the $BeF_2.4H_2O$ so produced is hydrolysed during dehydration. It can be made[2] by the direct action of gaseous hydrogen fluoride on beryllium oxide at 220° but it is usually obtained by Lebeau's method in which ammonium fluoroberyllate, $(NH_4)_2BeF_4$, is thermally decomposed[3]. Ammonium fluoroberyllate is made by dissolving beryllium hydroxide in aqueous ammonium bifluoride, and evaporating the solution. Since an industrial method of making beryllium is based on the thermal decomposition of ammonium fluoroberyllate, this has received considerable attention. Thilo and Schroder[4] believe the decomposition to take place in stages:

$$(NH_4)_2BeF_4 = NH_4BeF_3 + NH_4F \ (230-260°),$$

$$NH_4BeF_3 = BeF_2 + NH_4F \ (>270°)$$

Novoselova et al.[5] however, state that when ammonium fluoroberyllate is slowly and steadily raised in temperature decomposition occurs between 218–232° without the formation of intermediate phases. But when maintained at 240° the material decomposes by way of two intermediate ammonium compounds:

$$(NH_4)_2BeF_4 \rightarrow NH_4BeF_3 \rightarrow (NH_4)_2Be_2F_5 \rightarrow BeF_2.$$

At temperatures below 240° however, the reaction product retains some of the intermediate phases and the yield of beryllium fluoride is poor. The crystalline form of the beryllium fluoride produced, whether cristobalite or quartz type, also depends upon the conditions[5-7].

2. Properties of beryllium fluoride

An important aspect of the chemistry of beryllium fluoride and the fluoroberyllates is the structural relationship between these compounds and silica and the silicates respectively [8, 9, 39]. This is a particular instance of a general relationship between oxide and fluoride systems, arising from the similar radii (1.40 and 1.36Å)[10] and polarizabilities of the O^{2-} and F^- ions. Thus fluorides can act as weakened models of oxide systems provided the corresponding cations in the two structures have sufficiently similar radii and polarizabilities, and the charge on the cation in the fluoride structure is one half that in the oxide. For example, calcium fluoride and thoria are based on the same model, both have the same crystal structure but calcium fluoride, as the weakened model, is lower in

References p. 59

melting point and hardness, and higher in solubility and chemical reactivity.

The weaker structure of fluoride compared with oxide systems may be explained, assuming bonding in both systems to be predominantly electrostatic, from the fact that for two ions with charges Z_1, Z_2 and radii r_1, r_2 the binding energy

$$E = Z_1 \times Z_2 \times e^2/(r_1 + r_2)^2$$

and shows that the binding energy between the ions increases with their valency and decreases with their size.

Like silica and germanium dioxide, beryllium fluoride possesses a radius ratio Be : F of ~0.3, a figure which Goldschmidt[39] believes will allow compounds of the AX_2 type to form glasses. Beryllium fluoride is rather difficult to crystallise and for many years was known only as a glass. The structure of the glass has been studied by Warren and Hill[14] using X-ray methods, their results suggest that beryllium fluoride has a random network structure in which beryllium atoms are surrounded tetrahedrally by four fluorine atoms and fluorine atoms by two beryllium atoms. Such a three-dimensional network structure is analogous to those in vitreous silica and germanium dioxide.

Another aspect of the structural relationship between beryllium fluoride and silica is the existence of polymorphic crystalline forms of beryllium fluoride analogous to the polymorphic forms of silica; the quartz and cristobalite forms of beryllium fluoride are well established[8,9]. Russian workers, whose results up to 1959 are summarised in ref. 11, consider there are other polymorphic forms of beryllium fluoride, their conclusions being summarised in the phase diagram Fig. 3. They believe that beryllium fluoride has a β-quartz form stable at room temperature (corresponding to the low temperature quartz form of silica), this is transformed into an α-quartz form at 220° (corresponding to the high temperature quartz form of silica). The α-quartz form when slowly heated is transformed into what is possibly a tridymite form at 420–450°. At ~ 680° this phase itself is converted into an α-cristobalite form of beryllium fluoride, corresponding to the high temperature

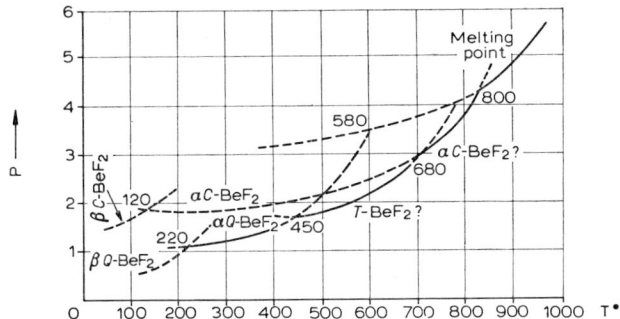

Fig. 3. Phase equilibrium diagram for beryllium fluoride.

cristobalite form of silica. The α-cristobalite modification of beryllium fluoride is said by Novoselova[11] to melt at 800°, in good agreement with the figure, 797°, given in ref. 18. A comparison of the melting point with that of silica (1710°) illustrates the relatively weakened nature of the beryllium fluoride structure.

The α-quartz form of beryllium fluoride has been said to melt at ~ 580°[11], when rapidly heated, although a figure of 540–548° appears more reliable[8, 9, 12, 13]. Significantly there is general agreement that the reconstructive transformations between the three main forms of beryllium fluoride are very sluggish, just as are similar transformations between the different forms of silica. Presumably this is because transitions, such as those between the quartz and tridymite forms of beryllium fluoride involve the breaking of F–Be–F bonds and new linking of BeF_4 tetrahedra. Such rearrangements in the solid state are normally slow processes, and, as with silica, the beryllium fluoride transformations are catalysed by small quantities of impurities.

The beryllium fluoride-silica relationship is further illustrated by the properties of the molten fluoride. It has a high viscosity[12], higher than that of most molten halides by a factor of ~ 10^6, and a low electrical conductivity[12, 15]. Both properties are considered to indicate that liquid beryllium fluoride possesses a non-ionic network structure similar to that of molten silicon dioxide or diboron trioxide.

References p. 59

Beryllium fluoride begins to vaporize at $\sim 800°$; its vapour pressure[16, 17] is ~ 100 mm at $1000°$, and its extrapolated boiling[16] point $1159°$. It is not dimeric in the vapour phase, and shows considerable thermal stability being less than 50% dissociated at $3000°$K. and 2×10^{-6} atmospheres[18]. Electron diffraction by the vapour[19] indicates a Be–F distance of 1.43Å, and emision infra-red spectroscopy measurements at $1000°$ show that the Be–F bond is strongly directional, at least in the relatively isolated environement of the vapour phase[20] (for a theoretical discussion of the latter result see ref. 21). The bonds have these directional properties despite the high proportion of ionic character which they possess (p. 38).

The sublimation of beryllium fluoride at $1038°$ (in presence of 2% beryllium metal and at 1 mm. pressure) has been used for its purification[22], and spectrographic analysis of the sublimate showed the following impurities to be present, Al 15, Cu 65, Si 14, Fe 150, Ca 50 and Mg 65 ppm. Attempts to employ this procedure on a large scale did not afford so pure a product.

3. Aqueous solutions of beryllium fluoride

Beryllium fluoride is a hygroscopic material with a very high solubility in, and affinity for, water, although its rate of dissolution is slow. The contrasting solubility of beryllium fluoride and silica is another illustration of the weaker bonding in fluoride systems. As the solubility of beryllium fluoride in water is limited to ~ 25 moles per l. at $25°$, which corresponds to two moles of water per mole of beryllium fluoride, it has been suggested[23] that the complex

$$\begin{bmatrix} H_2O & & F \\ & Be & \\ H_2O & & F \end{bmatrix}^0$$

is present in the solution.

Although the presence of the complex is consistent with beryllium fluoride being a weak electrolyte in water, as is shown by the

3   AQUEOUS SOLUTIONS OF BERYLLIUM FLUORIDE

low electrical conductivity and small molal freezing point depressions of the solutions, the actual nature of these solutions is more complicated. As shown by Prytz[24], using EMF and conductivity methods, autocomplex formation occurs; more recent measurements have disclosed the presence of BeF^{+}[25], BeF_2, BeF_3^- and BeF_4^{2-} ions in these solutions. The relative concentrations of the ionic species depends upon the F/Be ratio and the pH. Kleiner[26], from measurements of the optical density of the colour of the $FeSCN^{2+}$ ion in the system $Fe(NO_3)_3$–$KSCN$–NaF–$Be(NO_3)_2$, showed that BeF^+ was detectable at sodium fluoride concentrations of $\sim 10^{-4}M$. He quoted the following consecutive reaction constants, $\log_{10} Kn$, defined by

$$Kn = \frac{[BeF_n][H^+]}{[BeF_{n-1}][HF]},$$

viz. $\log K_1 = 2.89$, $K_2 = 1.94$, $K_3 = 0.56$, $K_4 = -1.00$. Other values for $\log K_1$, K_2 and K_3 are 2.12, 0.83 and 0.03 in $0.5M$ $NaClO_4$ by Yates[26a], and 1.99, 1.12 and 0.38 in $2M$–$HClO_4$ by Hardy, Greenfield and Scargill[26b].

Although claims have been made that hydroxy fluoroberyllate anions exist in alkaline solutions of beryllium fluoride[27], these claims are subject to dispute[28]. It appears that addition of alkali to a beryllium fluoride solution causes the precipitation of beryllium hydroxide only, although the shape of the pH-titration curve is dependent upon the F/Be ratio in solution.

4. Fluoroberyllates prepared from aqueous solution

As would be expected from the presence of BeF_3^- and BeF_4^{2-} ions in solution, salts of these anions can be crystallised under the correct conditions. Those of the latter are most readily obtained, but phase equilibria studies, in particular those of Novoselova and co-workers, show that fluoroberyllates containing a greater proportion of beryllium can be crystallised out from more concentrated solutions when the ratio F/Be 4 : 1. In the NH_4F–BeF_2–H_2O

system[29] $(NH_4)_2BeF_4$ and NH_4BeF_3 occur, and in the NaF–BeF$_2$–H$_2$O system[30] Na_2BeF_4, $NaBeF_3$ and $4NaF.3BeF_2.2H_2O$ have been identified. The identity of the last, a rather unusual compound, has been confirmed by X-ray methods. The compounds K_2BeF_4, $KBeF_3$ and $K_2Be_2F_5$ have been isolated[31], and also analogous compounds of rubidium[32].

The similarity between the BeF_4^{2-} type fluoroberyllates obtained from solution and the corresponding sulphates has been established by Ray[33]. Thus, the Co^{II}, Ni^{II} and Zn^{II} salts crystallize with six, six and seven molecules of water of crystallization respectively, as do the corresponding sulphates with which these hydrates are isomorphous. The fluoroberyllates also form double salts with the alkali metal fluoroberyllates, e.g. $M_2BeF_4.M^{II}BeF_4.6H_2O$, as do the corresponding sulphates. In general, the solubilities of salts of the BeF_4^{2-} ion are very similar to those of the sulphates although there are exceptions; thus Ag_2BeF_4 is very soluble and silver sulphate relatively insoluble. These similarities between the fluoroberyllates and the sulphates appears to be due to the similar size of the BeF_4^{2-} and SO_4^{2-} anions. Ray's conclusions have been recently confirmed by Rollier[34] who showed by X-ray methods that $BaBeF_4$ (crystallized from solution) is isomorphous with barium sulphate.

The preparation of the BeF_4^{2-} fluoroberyllates has also been studied by Perfect[35], who showed that the alkali and alkaline earth salts are readily obtained by adding the appropriate nitrate or chloride to a hot, nearly saturated solution of $(NH_4)_2BeF_4$. He claims that $MgBeF_4$ can be prepared in this way provided the salt is not kept too long in the boiling solution. This slowly and quantitatively decomposes it into beryllium fluoride and insoluble magnesium fluoride. A similar behaviour is shown by $CaBeF_4$. This observation of Perfect is of interest as the anhydrous MgF_2–BeF_2 system has not yet yielded a fluoroberyllate[36, 37].

Claims have been made that oxide fluoroberyllates result from the hydrolysis of the corresponding fluoroberyllate solutions[38]; instances cited are $CuBeF_4.4Cu(OH)_2.8H_2O$ and $NiBeF_4.4Ni(OH)_2.5H_2O$, but the findings await confirmation.

5. Anhydrous beryllium fluoride systems

As stated previously, there is a structural similarity between fluoroberyllates and silicates. A good illustration of this is provided by the $NaF - BeF_2$ system, which is a weakened model of the important $CaO-SiO_2$ system. The $NaF-BeF_2$ system has been well studied[4, 8, 9, 40-43], the American results[8, 9] appearing to be the most reliable and complete. The phase diagram Fig. 4, is very like that of the $CaO-SiO_2$ system[44]. The strong similarity of the two systems is emphasised by evidence for Na_3BeF_5 and $Na_3Be_2F_7$ in the $NaF-BeF_2$ system obtained by Novoselova et al.[45].

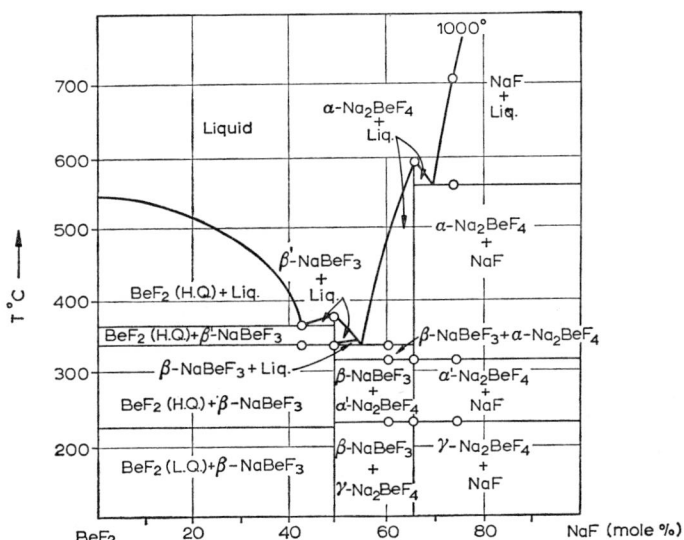

Fig. 4. Phase equilibrium diagram for the system $NaF - BeF_2$.
(Reproduced, with permission, from *J. Amer. Ceram. Soc.*, 36 (1953) 185.)

The most serious limitation of the fluoride model relationship is that it is not always possible to predict which of several fluoride systems will be a model for a particular oxide system or visa-versa[46]. For example, there is some doubt as to which particular silicate Li_2BeF_4 is related. Thilo and Lehmann[47] suggest fosterite (Mg_2

References p. 59

SiO$_4$), whilst Roy, Roy and Osborn[46] and Hahn[48] consider it to resemble willemite (Zn$_2$SiO$_4$) and phenacite (Be$_2$SiO$_4$), although the American workers admit that Li$_2$BeF$_4$ shows some similarities to fosterite. If Li$_2$BeF$_4$ is related to phenacite then it should not exhibit any polymorphic changes as none occur in the BeO–SiO$_2$ system[49]. Although Roy, Roy and Osborn consider Li$_2$BeF$_4$ not to show enantiotropy, Novoselova et al.[50] state that three polymorphic forms exist:

$$\alpha Li_2BeF_4 \overset{178°}{\rightleftharpoons} \beta Li_2BeF_4 \overset{360°}{\rightleftharpoons} \gamma Li_2BeF_4$$

The above discrepencies highlight the fact that, despite the various investigations, many of the beryllium fluoride systems described in the literature are probably incompletely characterised. One cause of this may be the sensitivity of such systems to moisture, with resulting hydrolysis to give oxygen containing beryllium compounds. Another typical difficulty, discussed by Roy, Roy and Osborn for the LiF–BeF$_2$ system[46], is that some compounds, such as LiBeF$_3$, have only a subsolidus stability range, so that their existence can easily be missed. In this particular example the subsolidus compound could be readily obtained by devitrifying a glass of its own composition.

It is of interest briefly to review the various suggestions which have been made concerning the model relationships thought to exist between various fluoroberyllate and silicate systems. Hahn[48] suggests that, in addition to the systems already mentioned, LiBeF$_3$ is related to the pyroxene group and NaLiBe$_2$F$_6$ to diopside. Jahn[51] considers that the compound Na$_3$LiBe$_2$F$_8$ is related to merwinite (MgO.3CaO.2SiO$_2$), whilst Toropov and Shchetnikova[52] suggest that Na$_2$KBe$_3$F$_9$ is similar to 2CaO.BaO.3SiO$_2$. Fluoride systems are related only to those silicates which possess either discrete tetrahedral silicate groups or silicate chains, and no fluoride analogue of sheet or framework silicates exist. Counts, Roy and Osborn[37] have shown that the MgF$_2$–BeF$_2$ system is similar to the TiO$_2$–SiO$_2$ system, especially in the lack of compound formation. (cf. Perfect, p. 44). Similarly, the CaF$_2$–BeF$_2$ system is a model for the ZrO$_2$–SiO$_2$ and ThO$_2$–SiO$_2$ systems in that only one compound is formed, $CaBeF_4$, which possesses an incongrunt melting point as do $ZrSiO_4$ and $ThSiO_4$.

Amongst other beryllium fluoride systems investigated has been the $RbF-BeF_2$ system[53, 54], which shows at least partial resemblences to the $BaO-SiO_2$ system. The compounds $RbBe_2F_5$, Rb_3BeF_5, $RbBeF_3$ and Rb_2BeF_4 were observed, the latter two compounds existing in polymorphic forms[53]. The analogous four compounds are also observed in the $CsF-BeF_2$ system[55, 56], all four compounds existing in polymorphic forms. In general, the $M_2^I BeF_4$ compounds are the most stable of the fluoroberyllates[55].

The tendency to possess numerous polymorphic forms is one of the most characteristic features of crystalline fluoroberyllates. One result of this is that the strong similarity between sulphates and the salts of the BeF_4^{2-} ion discussed by Ray[33] (p. 44) only strictly applies to that crystal form of the fluoroberyllate which is stable at room temperature. Thus $BaBeF_4$ has three modifications[57]:

$$\alpha BaBeF_4 \underset{}{\overset{360°}{\rightleftharpoons}} \beta BaBeF_4 \underset{}{\overset{870°}{\rightleftharpoons}} \gamma BaBeF_4,$$

and it is the α form which crystallizes in the rhombic lattice and which is isostructural with barium sulphate. The cell dimensions for α $BaBeF_4$ prepared under anhydrous conditions (a = 8.87, b = 5.30 and c = 7.02Å)[57] are in fair agreement with those reported by Rollier[34] for the material obtained from aqueous solution (a = 8.73, b = 5.65 and c = 6.61Å). Analysis of the SrF_2-BeF_2 system shows that $SrBeF_4$ also exists in a number of polymorphic forms[58]. At present nothing is known concerning the existence of polymorphic modifications of the other types of BeF_4^{2-} fluoroberyllates studied by Ray[33].

This account of beryllium fluoride would be incomplete without a short summary of American work on molten fluoride systems undertaken in conjunction with the development of a molten salt reactor for atomic energy purposes. Amongst the systems studied has been $NaF-BeF_2-ThF_4$ and $NaF-BeF_2-UF_4$[59], in which the compounds $NaF.BeF_2.ThF_4$ and $NaF.BeF_2.UF_4$ were observed, whilst in the $NaF-LiF-BeF_2$ system[54] the compounds $NaLiBe_3F_8$, $Na_5LiBe_3F_{12}$ and Na_2LiBeF_5 were identified.

The importance of trace quantities of oxygen in fluoride melts has been demonstrated by the American workers[60]. Addition of beryllium oxide to a fluoride melt can cause precipitation of certain of the components of the fluoride mixture as oxides for instance

$$2CeF_3 + 3BeO \rightleftharpoons Ce_2O_3 + 3BeF_2$$

Such reactions could be of interest as possible methods for removing reactor poisons from the molten fluoride fuel, although it must be noted that beryllium oxide itself is an undesirable impurity. In these fluoride melts the order of affinity for oxygen is $UO_2 \simeq ZrO_2 \rangle BeO \rangle ThO_2 \rangle Ce_2O_3$. Thus it is possible to remove uranium from a molten fluoride mixture by passing it down a column of beryllium oxide, UO_2 is precipitated and tenaciously held on the beryllium oxide surface. Use of thirty grams of beryllium oxide per kilogram of molten salt mixture (LiF, BeF_2 and ThF_4) reduces the uranium in solution from 1800 p.p.m. to less than one p.p.m.[60, 61].

Another illustration of the reactivity of molten beryllium fluoride systems towards oxygen is shown by the formation of single crystals of beryllium oxide up to 200µ in size by the action of water vapour on a lithium fluoride rich $LiF-BeF_2$ melt[62] at 800°. As the concentration of beryllium fluoride in the mixture decreased so did the size of the beryllium oxide crystals obtained, although any size reduction is accompanied by an increase in purity. Total impurities were ca. 500–700 p.p.m., of which $\sim 80\%$ were attributable to solvent and container corrosion products.

There is some evidence that fluoroberyllate species can occur in the vapour as well as in the liquid and solid phases. Thus vapour pressure measurements[63] on the $NaF-BeF_2$ system between 509–1061° are considered to show that both $NaBeF_3$ and Na_2BeF_4 co-exist with BeF_2 and NaF in the vapour phase.

6. Beryllium fluoride glasses

Beryllium fluoride systems readily form glasses, as might be expected from their similarity to silicates. Indeed, the difficulty with beryllium fluoride systems often is to induce crystallization rather than the establishment of the correct conditions for obtaining the glassy state. Glasses are more readily obtained the greater the proportion of beryllium fluoride in the system, although large quantities of beryllium fluoride cause the glasses to be hygroscopic and difficult to handle experimentally (*e.g.* ref. 64). In order to minimize hydrolysis, it is best to undertake all operations in a water free hydrogen fluoride atmosphere. Failure to eliminate moisture results in turbid glasses contaminated with beryllium oxide[65, 66].

An interesting illustration of the weakened structure of beryllium fluoride as compared to silicate systems is given by Zarzycki and Naudin[67], who showed from infrared spectroscopic measurements on glassy beryllium fluoride and silica, that the force constants of the Be–F and Si–O bonds are respectively 1.1×10^5 and 4.0×10^5 dynes per cm. The lower force constant of the Be–F bond measures the relatively weakened nature of the beryllium fluoride systems. It was also observed that, in addition to the two significant beryllium fluoride bands at 14 and 16μ, two bands due to oxygen impurity were observed at 10.5 and 11.5μ[67]. This further illustrates the tendency of anhydrous beryllium fluoride systems to oxygen contamination.

7. *Miscellaneous reactions of beryllium fluoride*

The solubility of beryllium fluoride in anhydrous hydrofluoric acid is 0.015 at 11° and 0.014 at −24° g per 100 g solvent[38]. As these values are rather greater than the solubilities of beryllium hydroxide in water, they indicate that beryllium fluoride is a stronger base in anhydrous hydrofluoric acid than is beryllium hydroxide in water.

With ammonia, beryllium fluoride forms at −78.5° an ammine, $BeF_2 \cdot NH_3$, which decomposes at higher temperatures[69]. The formation by beryllium fluoride of only a single unstable ammine, is in contrast to the behaviour of the other three beryllium halides.

References p. 59

8. Preparation of beryllium chloride

Beryllium chloride is prepared by dry methods, due to the ease with which the tetrahydrate undergoes hydrolysis during dehydration. (cf. ref. 70). One of the most convenient preparative methods is to heat a mixture of beryllium oxide and carbon in chlorine:

$$BeO + Cl_2 + C \overset{600-800°}{=} BeCl_2 + CO$$

Alternatively, the action of sulphur monochloride or phosgene on beryllium oxide, or the direct chlorination of beryllium carbide (at 800°) can be used. Beryllium chloride is not conveniently made by direct hydrochlorination of beryllium oxide as the reaction is reversible:

$$BeO + 2HCl \rightleftharpoons BeCl_2 + H_2O,$$

the equilibrium lying over on the left except at very high temperature[74, 75].

9. Properties of beryllium chloride

Beryllium chloride has variously reported melting points of 399°[76], 404°[72], 405°[76], 416°[77], and 425°[78], and boiling points of 482°[106], 488°[76] and 520°[74, 78]; it is impossible at present to suggest which are the correct values. Beryllium chloride is soluble in water and in organic solvents. On sublimation (350-380° at pressures of 20–50μ) it forms fibrous crystals[106] reminiscent of cotton wool, which are also obtained on rapid cooling of molten beryllium chloride. Novoselova et al.[78] consider this form to be a metastable phase (α′ modification). X-ray studies[79] have shown this α′ modification to be orthorhombic, a = 9.86, b = 5.36, c = 5.22Å and z = 4, isomorphous with dimethylberyllium and silicon disulphide. It consists of continuous chains of distorted $BeCl_4$ tetrahedra linked together by opposite edges and situated parallel to the c axis. The bond angles indicate that all the bonds contain an electron pair, so that the chlorine atoms must use an unshared pair of electrons to form the chlorine bridge bonds. Thus beryllium chloride can be considered to contain different type bonding compared to electron deficient molecules such as dimethylberyllium (see p. 92).

On heating the α' modification to 250° it changes to a cubic modification (β'), a = 8.27Å and z = 8, which itself changes to the stable β phase at 340°[78]. The latter phase, like the α' phase is orthorhombic, a = 16.08, b = 14.48, c = 10.10Å and z = 36. The β phase appears to have essentially the same structure as the high temperature form of beryllium iodide[80]. Slow cooling of molten beryllium chloride causes the crystallization of a fourth phase (α) at 425°, this phase transforms at 405° into the stable β phase. Little is known of the properties of this α phase[78].

The polymorphism of beryllium chloride arises from there being different ways of packing $BeCl_4$ tetrahedra into the crystal. The polymorphic transitions arising from such causes would give rise to only small heat changes, and hence be characterised by only slow rates of interconversion. The variation in the melting point of beryllium chloride noted above is probably a consequence of the existence of a number of polymorphic forms which only slowly come into equilibrium.

Molten beryllium chloride has a very low electrical conductivity[81-83]. The degree of dissociation of beryllium chloride just above its melting point is 0.0086 for the reaction[84]:

$$BeCl_2 \rightleftharpoons Be^{2+} + 2Cl^-,$$

and 0.28 for the reaction:

$$BeCl_2 \rightleftharpoons BeCl^+ + Cl^-$$

Beryllium chloride is associated in the vapour phase in the range 500–600°; Rahlfs and Fischer[76] state that 23% of the beryllium in the vapour phase is present as Be_2Cl_4 dimers at 564°. The presence of dimers in this temperature range has been confirmed by Buchler and Klemperer[20] by infrared methods, these authors suggest the structure

$$Cl - Be \genfrac{}{}{0pt}{}{\diagup Cl \diagdown}{\diagdown Cl \diagup} Be - Cl$$

for the dimer. At higher temperatures dissociation of the dimer to give monomeric $BeCl_2$ occurs, dissociation is complete at 1000°[18]. Electron diffraction measurements[19, 85] show the monomer to be linear, with a Be–Cl distance of 1.77Å (*cf.* the value of 1.70Å for the Be–Cl distance obtained from spectroscopic data[86]). The Be–Cl bond length in beryllium chloride vapour is shorter than in the solid (2.02Å)[79]; this difference is compatible with the different bonding involved, *i.e.* sp^3 in the solid and sp in the vapour.

10. Aqueous solutions of beryllium chloride

The marked difference in properties between beryllium chloride and fluoride, a difference shown up just by considering the properties of the pure compounds, becomes even more evident when the chemical properties of these two compounds are considered.

Beryllium is a member of that group of elements, the stabilities of whose halide complexes lies in the order $F \gg Cl \rangle Br \rangle I$[87, 88]. The bonding in the halide complexes of these elements is essentially electrostatic and the strongest bonds will be formed by ligands combining a small ionic radius with a low polarizability, parameters which decrease monotonically from I^- to F^-. The relatively greater stability of the fluoride complexes arises from the very small size of the fluoride ion relative to that of the other halides (radii of F^-, Cl^-, Br^- and I^- are 1.36, 1.81, 1.95 and 2.16Å respectively[1]). An immediate illustration of the lower stability of chloride complexes of beryllium is given by the fact that, unlike beryllium fluoride, beryllium chloride is a strong electrolyte in aqueous solution[89].

Although beryllium chloride, like beryllium fluoride, is very soluble in water, no anionic chlorocomplexes of beryllium have been established as existing in solution, except possibly under extreme conditions. Thus Prytz[24] has shown by potentiometric methods that no beryllium chlorocomplexes occur in $BeCl_2$–KCl solutions. Kraus *et al.*[90] find that beryllium chloride is taken up by an anion-exchange resin from 12M lithium chloride solution, and Ohtaki and Yamasaki[91] obtained evidence from ion-exchange and spectro-photometric measurements that a cationic beryllium chloride complex exists in concentrated solutions of beryllium chloride and hydrochloric acid. A value for log K of -0.66 for formation constant of the $BeCl^+$ complex was obtained by Hardy et al.[26b] from solvent extraction data.

Despite much effort, all attempts to prepare chloroberyllates from aqueous solution have failed, $BeCl_2.4H_2O$ and the other metal halide were always the only solid phases isolated. The systems studied were: $BeCl_2-HCl-H_2O$[92], $BeCl_2-M^ICl-H_2O$ (where M^I = NH_4 and the alkali metals)[93-95], $BeCl_2-M^{II}Cl_2-H_2O$ (where M^{II} = Mg, Ba, Zn and Hg)[97, 98], and $BeCl_2-AlCl_3-H_2O$[96].

An apparent exception to the above finding is provided by the beryllium chloro- bromo- and iodo-mercurates prepared by Slavvo[99]. If mercuric chloride is added to a solution of beryllium chloride in concentrated hydrochloric acid $BeCl_2.HgCl_2.5H_2O$ is precipitated, whilst under other conditions $Be(OH)Cl.HgCl_2.2H_2O$ is obtained. It has been suggested[100] that the latter compound should be formulated as $[BeOH]_3[HgCl_3]_3.6H_2O$ (see p. 10) and, as Slavvo considers that $BeCl_2.HgCl_2.5H_2O$ should be formulated as $Be[HgCl_4].5H_2O$, these two compounds cannot be considered as halogenoberyllates.

11. Anhydrous beryllium chloride systems

Liquid beryllium chloride itself is a poor electrical conductor, but addition of even small quantities of alkali metal chlorides considerably increases the conductivity. Delimarski and co-workers[101-104], by means of electrical conductivity and EMF measurements, have obtained evidence for both anionic and cationic beryllium species in $NaCl-BeCl_2$ melts depending upon the relative concentrations of the components. Schmidt[72] showed that chloroberyllates $M_2^I BeCl_4$ (M^I = Na, Li and Tl^I) could be isolated from such fused salt mixtures under the correct conditions. Thus there is good evidence that chloroberyllates can exist in anhydrous systems, although they are clearly less stable than the fluoroberyllates.

An interesting claim has been made by Smirnov and Chukreev[105] that at low current densities a beryllium anode will dissolve in a chloride melt (LiCl–KCl eutectic mixture + 10% $BeCl_2$) to give beryllium of charge number one:

$$Be^{2+}{}_{(melt)} + Be = 2Be^{+}{}_{(melt)}$$

The salt mixture containing Be^+ is outwardly similar to the original mixture but evolves hydrogen on addition of water. These authors find that:

$$K = \frac{[Be^+]^2}{[Be^{2+}]} = (2.36 - \frac{4904}{T}) \pm 0.10,$$

this equation showing that the quantity of Be^+ in equilibrium with Be^{2+} increases with temperature; the equilibrium concentration of Be^+ at room temperature being quite negligible ($K \approx 4.2 \times 10^{-15}$ at 25°). However, in view of the number of occasions where claims for the production of unusual valency states in fused salt systems have proved erroneous, these observations of the Russian workers are in need of confirmation.

Electrolysis of fused $NaCl$–$BeCl_2$ mixtures with a mercury cathode under an inert atmosphere affords beryllium amalgam[107]. This material is unstable in air and decomposes spontaneously to give a black powder.

12. *Beryllium chloride complexes with neutral ligands*

The greater range and stability of the complexes of beryllium chloride with neutral ligands, compared to those formed by beryllium fluoride, is illustrated by the ammines. Whilst beryllium fluoride forms only one unstable ammine (see p. 49), beryllium chloride forms four *viz.* $BeCl_2.12NH_3$, $BeCl_2.6NH_3$, $BeCl_2.4NH_3$ and $BeCl_2.2NH_3$[69, 108, 109]. The lower ammines are relatively stable, *e.g.* $BeCl_2.4NH_3$ has a vapour pressure of only 6 mm at 156°. The diammine of beryllium chloride is the most difficult to prepare, as the vigorous conditions required to remove ammonia from $BeCl_2$. $.4NH_3$ causes partial decomposition.

Numerous complexes of the type $BeCl_2X_2$, where X represents a wide variety of organic ligands, have been prepared, either by direct interaction or by addition of the ligand to an ethereal solution of beryllium chloride. Examples of such complexes are those with pyridine[110], acetone[110], nitriles[111], aldehydes[111], quinoline[112], aliphatic amines[113], piperidine[114], thiourea[115] and tetrahydrofuran[116]. Sometimes the stoichiometry of these compounds departs from the usual 1 : 2 beryllium : ligand relation, thus methylamine forms a tetrammine[111] analogous to $BeCl_2.4NH_3$. All of these complexes, as well as the ammines, are decomposed by water and must be prepared under anhydrous conditions.

The most important of the beryllium chloride addition complexes is the dietherate, $BeCl_2.2Et_2O$. It is obtained by dissolving anhydrous beryllium chloride in dry ether; two liquid layers are obtained, the top layer being a solution of $BeCl_2.2Et_2O$ in ether whilst the lower layer is a solution of ether in $BeCl_2.2Et_2O$, from which the compound is crystallised. Its melting point has been reported as 33° by Fricke and Robke[113] and 43–44.5° by Nespital[117]. Novoselova et al.[118] say that $BeCl_2.2Et_2O$ prepared by the above method is hexagonal (a = 7.65, c = 20.00Å, z = 3), and has a melting point of 43°. These crystallographic data were obtained by single crystal measurements on material prepared immediately before making measurements. Powder photograph data did not agree with the above observation, and it appears that the hexagonal form of $BeCl_2.2Et_2O$ is converted by grinding into a new stable form. This stable form has markedly different properties to the hexagonal form e.g. it is almost insoluble in ether and decomposes without melting when heated. Unless otherwise stated all references to $BeCl_2.2Et_2O$ elsewhere in this book will be confined to the better known hexagonal form.

Unlike anhydrous beryllium chloride, the dietherate is soluble in solvents such as benzene or carbon tetrachloride. Silber[119] showed by ebullioscopic methods that $BeCl_2.2Et_2O$ is virtually undecomposed in boiling benzene. Ulich and Nespital[120] have found that the dietherate has a moment of 6.74D in benzene, which is unchanged on dilution showing that no dissociation of the complex takes place. Novoselova et al.[118] explain this large dipole moment of $BeCl_2.2Et_2O$ by suggesting that the addition of two ether molecules causes a change in the Cl–Be–Cl bond angle on passing from the linear $BeCl_2$ to the near tetrahedral $BeCl_2.2Et_2O$. Thus the observed moment is made up from the moment of the two ether molecules together with a contribution provided by the distortion of the $BeCl_2$ molecule.

Action of anhydrous hydrogen chloride on $BeCl_2.2Et_2O$ is stated to give a greenish liquid complex, $HBeCl_3.2Et_2O$[121]. Addition of ethereal pyridine to this complex affords (after strong cooling) crystals of $HBeCl_3.2pyridine$. On heating, this material melts at 108° but does not solidify again on cooling. After a few days the uncrystallised melt deposits crystals of the same composition as the original complex, but with a melting point raised to 145°. This change may be due to transfer of a molecule of pyridine from the outer to the inner coordination sphere, although it is possible that this change represents a polymorphic transition similar to that observed with $BeCl_2.2Et_2O$.

An unusual two stage reaction takes place between beryllium chloride and chloroform:

$$BeCl_2 + CHCl_3 = CCl_3-BeCl + HCl\ (20°),$$

$$Cl_3C-BeCl + CHCl_3 = Cl_3C-Be-CCl_3 + HCl\ (61°)$$

That this reaction occurs by simple contact of the reagents at room temperature illustrates the high chemical reactivity of anhydrous beryllium chloride.

13. Miscellaneous reactions of beryllium chloride

An important feature of the complex chemistry of beryllium chloride is its power to act as a Friedel Crafts catalyst[122, 123]. It is very similar to aluminium chloride in this respect, although higher temperatures are required when beryllium chloride rather than aluminium chloride is used as catalyst. Beryllium chloride can also act as a polymerization catalyst (*e.g.* polymerization of epoxides[124]).

14. Preparation and properties of beryllium bromide

Beryllium bromide is prepared by similar methods to those used for the chloride *e.g.* by action of bromine on beryllium metal (550°)[125] or beryllium carbide (500°)[72], and by action of bromine on a mixture of beryllium oxide and carbon at 1100–1200°[118]. It cannot be made by wet methods since the tetrahydrate hydrolyses on drying. Beryllium bromide melts at 488° and sublimes below this at 473°[76]. In its other properties it appears similar to the chloride although the quantity of published work is relatively limited.

15. Reactions of beryllium bromide

Beryllium bromide is readily soluble in water to give solutions containing the hydrated beryllium cation; no bromoberyllates have ever been reported, either in aqueous or non-aqueous media. Beryllium bromide is soluble in the more polar organic solvents (*e.g.* ethanol or pyridine[126]) and, like the chloride, it forms a large number of addition complexes of the $BeBr_2.2X$ type with neutral ligands. As well as a series of ammines[69, 108], beryllium bromide forms the compound $BeBr_2.2H_2S$ with hydrogen sulphide at $-78°$[127]; descriptions of its complexes with nitrogen or oxygen containing ligands are given in refs. 117, 120 and 125. Beryllium bromide can act as a Friedel Crafts catalyst, although less efficiently than beryllium chloride[128].

Beryllium bromide forms a dietherate, $BeBr_2.2Et_2O$, similar to the corresponding chloride compound[129]. This material crystallizes in the monoclinic system, melts at 53°, begins to decompose at 70° and is completely decomposed at 160–70°. As with the corresponding dietherate of beryllium chloride, $BeBr_2.2Et_2O$ forms two layers with ether, the lower layer is a solution of ether in the dietherate and the upper layer a solution of dietherate in ether. In contact with ether the melting point of $BeBr_2.2Et_2O$ is depressed to 37°[129], a fact which probably explains the variation of the melting point of this compound given in the literature[117, 125].

16. Preparation and properties of beryllium iodide

Beryllium iodide is prepared by action of hydrogen iodide on beryllium carbide at 700°[3,108], by heating beryllium oxide with the calculated amount of aluminium iodide in a sealed tube[130], or by heating iodine and beryllium metal in a sealed tube[125]. Care must be taken to exclude moisture during preparation as beryllium iodide is very easily hydrolysed by water, so easily that the tetrahydrate, $BeI_2.4H_2O$, cannot be prepared[131] (the ease of hydrolysis of the beryllium halides increases from the fluoride to the iodide).

Beryllium iodide exists in two polymorphic forms[80]. The tetragonel form (a = 6.12, b = 10.63Å and z = 4) is stable at room temperature, and is the form commonly encountered. The second modification is orthorhombic (a = 16.48, b = 16.70, c = 11.63Å and z = 32) and is stable only above 350°. The reasons for the polymorphism of beryllium iodide are probably similar to those accounting for polymorphism in beryllium chloride (see p. 50).

Beryllium iodide melts at 510° and boils at 590°[125] although these figures are subject to variation e.g. figures of 480° and 488° are quoted for the melting and boiling point in ref. 132. Beryllium iodide is relatively stable in the vapour phase and is not appreciably decomposed into its elements below 1200°[133], although Sloman[134] has reported preparing beryllium metal by thermal decomposition of beryllium iodide vapour on a hot metal filament, it is doubtful whether this method can in fact be applied[135]. Beryllium iodide is dimeric in the vapour phase at temperatures near to its melting point but it dissociates into the monomer as the temperature increases. The Be–I bond length in the monomer has been shown to be 2.12Å by electron diffraction measurements[19].

17. Reactions of beryllium iodide

Relatively little has been published on the reactions of beryllium iodide. Iodoberyllates are not known (for ammonobasic iodoberyllates see p. 20), but beryllium iodide does resemble the chloride and bromide in forming addition compounds with neutral ligands. The tendency for such compounds to be formed, and their subsequent stability, appears to decrease on passing from beryllium chloride to iodide. For example, $BeI_2.13NH_3$, $BeI_2.6NH_3$ and $BeI_2.4NH_3$ are known in the beryllium iodide-ammonia system[108], but beryllium iodide is only slightly soluble in ether, and no compound analogous to the dietherates of beryllium chloride or bromide appear to be formed[125].

REFERENCES

1 PAULING, L., *The Nature of the Chemical Bond*. 3rd ed., Cornell Univ. Press, Ithaca, N.Y., 1960
2 HYDE, K. R., O'CONNER, D. J. AND WAIT, E., *J. Inorg. Nucl. Chem.*, 6 (1958) 14.
3 LEBEAU, P., *Compt. Rend.*, 126 (1898) 1418.
4 THILO, E. AND SCHRODER, H., *Z. Phys. Chem.*, 197 (1951) 39.
5 BREUSOV, O. N., VAGUTOVA, N. M., NOVOSELOVA, A. V. AND SIMANOV, YU. P., *Zhur. Neorg. Khim.*, 4 (1959) 2213.
6 NOVOSELOVA, A. V., SIMANOV, YU. P., CHERNYKH, V. I. AND YARAMBASH, E. I., *ibid.*, 1 (1956) 2071.
7 NOVOSELOVA, A. V., BREUSOV, O. N., KIRKINA, D. F. AND SIMANOV, YU. P., *ibid.*, 1 (1956) 2670.
8 ROY, D. M., ROY, R. AND OSBORN, E. F., *J. Amer. Ceram. Soc.*, 33 (1950) 85.
9 *Idem, ibid.*, 36 (1953) 185.
10 WELLS, A. F., *Structural Inorganic Chemistry*, 3rd. ed., Oxford University Press (1961).
11 NOVOSELOVA, A. V., *Uspeckhi Khim.*, 28 (1959) 33; *U.S.A.E.C. report AEC-tr-3992*.
12 MACKENZIE, J. D., *J. Chem. Phys.*, 32 (1960) 1150.
13 THOMA, R. E., INSLEY, H., FRIEDMAN, H. A. AND WEAVER, C. F., *J. Phys. Chem.*, 64 (1960) 865.
14 WARREN, B. E. AND HILL, C. F., *Z. Krist.*, 89 (1934) 484.
15 NEUMANN, B. AND RICHTER, H., *Z. Electrochem.*, 31 (1925) 484.
16 SENSE, K. A., SNYDER, M. J. AND CLEGG, J. W., *J. Phys. Chem.*, 58 (1954) 223.

17 KHANDAMIROVA, N. E. et al., Zhur. Neorg. Khim., 4 (1959) 2192.
18 BREWER, L. M., in QUILL, L. L., The Chemistry and Metallurgy of Miscellaneous Materials, McGraw Hill, New York (1950) pp. 197 and 215.
19 AKISHIN, P. A., SPIRIDONOV, V. P. AND SOBOLOV, G. A., Dokl. Akad. Nauk S.S.S.R., 118 (1958) 1134.
20 BUCHLER, A. AND KLEMPERER, W. J., J. Chem. Phys., 29 (1958) 121.
21 BARRY, R. S., ibid., 30 (1959) 286.
22 Experiments undertaken by the Brush Beryllium Co., quoted by MOORE, R. E., U.S.A.E.C. report ORNL-2938 (1960).
23 LINNELL, R. H. AND HAENDLER, H. H., J. Phys. Chem., 52 (1948) 819.
24 PRYTZ, M., Z. Anorg. Chem., 231 (1937) 238.
25 TANANAEV, V. AND DEICHMAN, E. N., Bull. Acad. Sci. U.R.S.S. Classe Sci. Chim., (1947) 591; Izvest. Akad. Nauk S.S.S.R. Otdil Khim. Nauk, (1949) 144; ibid., (1951) 26; TANANAEV, V. AND VINOGRADOVA, A. D., Zhur. Neorg. Khim., 5 (1960) 321.
26 KLEINER, K. E., Zhur. Obsch. Khim., 21 (1951) 18.
26a YATES, L. M., Thesis, State Coll., Washington, 1955, Univ. Microfilms 15662.
26b HARDY, C. J., GREENFIELD, B. F. AND SCARGILL, D., U.K.A.E.A. Report AERE-R 3316 (1960); idem, J. Chem. Soc., (1961) 174.
27 MITRA, G., J. Ind. Chem. Soc., 32 (1955) 241.
28 GUPTA, A. K. S., ibid., 33 (1956) 433; Sci. and Culture, 23 (1958) 492.
29 NOVOSELOVA, A. V. AND AVERKOVA, M. YU., Zhur. Obsch. Khim., 9 (1939) 1063.
30 VOROB'EVA, O. I., NOVOSELOVA, A. V., ZHASMIN, A. G. AND SIMONOV, YU. P., Zhur. Neorg. Khim., 1 (1956) 516.
31 NOVOSELOVA, A. V., TAMM, N. S. AND VOROB'EVA, O. I., Khim. Redkikh Elementov, Akad. Nauk S.S.S.R. Inst. Obshchei. Neorg. Khim., No. 2 (1955) 3; Chem. Abs. 50 (1956) 3057.
32 TAMM, N. S. AND NOVOSELOVA, A. V., Zhur. Neorg. Khim., 2 (1957) 1428.
33 RAY, N. N., Z. Anorg. Chem., 201 (1931) 289; ibid., 205 (1932) 257; ibid., 206 (1932) 209; ibid., 227 (1936) 32 and 103; ibid., 241 (1939) 165.
34 ROLLIER, M. A., Gazz. Chim. Ital., 84 (1954) 663.
35 PERFECT, F. H., Proc. Penn. Acad. Sci., 26 (1952) 54.
36 VENTURELLO, G., Atti Reale Accad. Sci. Torino, 76 (1941) 559.
37 COUNTS, W. E., ROY, R. AND OSBORN, E. F., J. Amer. Ceram. Soc., 36 (1953) 14.
38 MITRA, G. AND RAY, N. N., J. Ind. Chem. Soc., 32 (1955) 43.
39 GOLDSCHMIDT, V. M., Skrifter Norske Videnkaps. Akad. Oslo Mat. Naturv. Klasse, 6 (1926) 104.
40 NOVOSELOVA, A. V., LEVINA, M. E., SIMANOV, YU. P., AND ZHASMIN, A. G., Zhur. Obsch. Khim., 14 (1944) 385.
41 O'DANIEL, H. AND TSCHEISCHWILI, L., Z. Krist, 103 (1941) 178; ibid., 104 (1942) 124.
42 JAHN, W. AND THILO, E., Z. Anorg. Chem., 274 (1953) 72.
43 THILO, E. AND LIEBAU, F., Z. Phys. Chem., 199 (1952) 125.
44 LEVINE, E. M., MCMURDIE, H. F. AND HALL, F. P., Phase Diagrams for Ceramists, The Amer. Ceram. Soc., Columbus, Ohio, 1956.

45 NOVOSELOVA, A. V. AND SIMANOV, YU. P., *Uchenye Zapiski Moskov Gosudarst Univ. im M. V. Lomonosova*, No. 174 (1955) 7; *Chem. Abs.*, 51 (1957) 5615.
46 ROY, D. M., ROY, R. AND OSBORN, E. F., *J. Amer. Ceram. Soc.*, 37 (1954) 300.
47 THILO, E. AND LEHMANN, H. A., *Z. Anorg. Chem.*, 258 (1949) 332.
48 HAHN, T., *Neues Jahrb. Mineral. Abhandl.*, 86 (1953) 1.
49 MORGAN, R. A. AND HUMMEL, F. H., *J. Amer. Ceram. Soc.*, 32 (1949) 255.
50 NOVOSELOVA, A. V., SIMANOV, YU. P. AND YAREMBASH, E. I., *Zhur. Fiz. Khim.*, 26 (1952) 1244.
51 JAHN, W., *Z. Anorg. Chem.*, 276 (1954) 3 and 113; *ibid.*, 277 (1954) 5 and 274.
52 TOROPOV, N.A. AND SHCHETNIKOVA, I. L., *Zhur. Neorg. Khim.* 2(1957)1855.
53 GREBENSHCHIKOV, R. G., *Dokl. Akad. Nauk S.S.S.R.* 114 (1957) 316.
54 *Phase Diagrams of Nuclear Reactor Materials*, THOMA, R. E. ed, *U.S.A.E.C. report ORNL-2548* (1959); LEVIN, E. M. AND MCMURDIE, H. F., *Phase Diagrams for Ceramists. Part II*. The Amer. Ceram. Soc., Columbus, Ohio, 1959.
55 BREUSOV, O. N., NOVOSELOVA, A. V. AND SIMANOV, YU. P., *Dokl. Akad. Nauk S.S.S.R.*, 118 (1958) 935.
56 BREUSOV, O. N. AND SIMANOV, YU. P., *Zhur. Neorg. Khim.*, 4 (1959) 2582.
57 KIRKINA, D. F. NOVOSELOVA, A. V. AND SIMANOV, YU. P., *Zhur. Neorg. Khim.*, 1 (1956) 125.
58 BREUSOV, O. N., TRAPP, G., NOVOSELOVA, A. V. AND SIMANOV, YU. P., *ibid.*, 4 (1959) 679.
59 THOMA, R. E., WEAVER, C. F., FRIEDMAN, H. A. AND HARRIS, L. A., *J. Amer. Ceram. Soc.*, 43 (1960) 608.
60 *U.S.A.E.C. report ORNL-2931* (1960).
61 SHAFFER, J. H., GRIMES, W. R., WATSON, G. M., CUNES, R., STRAIN, J. AND KELLY, M. J., *U.S.A.E.C. report TID-7160* (1960).
62 *U.S.A.E.C. report ORNL-3127* (1961); see also HARNS, W. O., in the *Proceedings of the beryllium oxide meeting*, Oak Ridge National Laboratory, 1960. *U.S.A.E.C. report TID-7602*.
63 SENSE, K. A., STONE, R. W. AND FILBERT, R. E., *U.S.A.E.C. report BMI-1186* (1957).
64 IMAAKA, M. AND MIZUSAUA, S., *J. Ceram. Soc. Japan*, 61 (1953) 13.
65 NOVOSELOVA, A. V., *Izvest. Vyshikh Ucheb. Zavedenii, Khim. Khim. Tekhnol.*, 2 (1959) 751; *Chem. Abs.*, 54 (1960) 7087.
66 VOGEL, W. AND GERTH, K., *Glastech. Ber.*, 31 (1958) 15.
67 ZARZYCKI, J. AND NAUDIN, F., *Verres et Refractories*, 14 (1960) 13.
68 JACKE, A. W. AND CADY, G. H., *J. Phys. Chem.*, 56 (1952) 1106.
69 BILTZ, W., KLATTE, K. A. AND RAHLFS, E., *Z. Anorg. Chem.*, 166 (1927) 351.
70 HECHT, H., *ibid.*, 254 (1947) 37.
71 FIELD, G. W., *J. Amer. Chem. Soc.*, 61 (1939) 1817.
72 SCHMIDT, J. M., *Bull. Soc. Chim. France*, 39 (1926) 1686; *Ann. Chim. France*, 11 (1929) 351.

73 TANNENBAUM, I. R., *Inorganic Syntheses*, 5 (1957) 22, McGraw Hill, New York, N.Y.
74 SPITZIN, V., *Z. Anorg. Chem.*, 189 (1930) 343.
75 BESSON, J., *Bull. Soc. Chim. France*, (1950) 1175.
76 RAHLFS, O. AND FISCHER, W., *Z. Anorg. Chem.*, 211 (1933) 349.
77 KLEMM, W., *Z. Anorg. Chem.*, 152 (1926) 243.
78 KUVYRKIN, O. N., BREUSOV, O. N., NOVOSELOVA, A. V. AND SEMENENKO, K. N., *Zhur. Fiz. Khim.*, 34 (1960) 343.
79 RUNDLE, R. E. AND LEWIS, P. N., *J. Chem. Phys.*, 20 (1952) 132.
80 STARITZKY, C. E., DOUGLAS, R. M. AND JONSON, R. E., *J. Amer. Chem. Soc.*, 79 (1957) 2037.
81 VOIGT, A. AND BILTZ, W., *Z. Anorg. Chem.*, 133 (1924) 280.
82 BILTZ, W. AND KLEMM, W., *ibid.*, 152 (1926) 268.
83 BJERRUM, N., *Ber.*, 62 (1929) 1091.
84 MARKOV, B. F. AND DELIMARSKI, YU. K., *Ukrain. Khim. Zhur.*, 19 (1953) 255; *Chem. Abs.*, 49 (1955) 3623.
85 AKISHIN, P. A. AND SPIRIDONOV, V. P., *Kristallografiya*, 2 (1957) 4 and 475.
86 HERZBERG, G., *Spectra of Diatomic Molecules*. D. van Nostrand, New York, N.Y. 1950.
87 AHRLAND, S. AND LARSON, R., *Acta Chem. Scand.*, 8 (1954) 354; AHRLAND, S., *ibid.*, 10 (1956) 723.
88 CARLESON, B. G. AND IRVING, H. M. N., *J. Chem. Soc.*, (1954) 4390.
89 HARNED, H. S. AND OWEN, R. B., *The Physical Chemistry of Electrolyte Solutions*. Reinhold, New York, N.Y. 1950.
90 KRAUS, K. A., NELSON, F., CLOUGH, F. B. AND CARLSTON, R. C., *J. Amer. Chem. Soc.*, 77 (1955) 1391.
91 OHTAKI, H. AND YAMASAKI, K., *Bull. Chem. Soc. Japan*, 31 (1958) 6.
92 LEIKINA, B. H. AND NOVOSELOVA, A. V., *Zhur. Obsch. Khim.*, 7 (1937) 241.
93 NOVOSELOVA, A. V. AND SOSNOVSKAYA, I. G., *ibid.*, 21 (1951) 813.
94 NOVOSELOVA, A. V., PASHINKIN, A. S. AND SEMENENKO, K. M., *Vestnik Moskov Univ.*, 10 (1955) 49; *Chem. Abs.*, 49 (1955) 11381.
95 BLINDIN, V. P., *Zhur. Obsch. Khim.*, 26 (1956) 1281.
96 Idem, *Zhur. Neorg. Khim.*, 1 (1956) 2633.
97 BLINDIN, V. P., GORDIENKO, V. I. AND SHALVEROVA, O. K., *ibid.*, 1 (1956) 2623.
98 BLINDIN, V. P., *ibid.*, 2 (1957) 1149.
99 SLAVVO, A. V., *Zhur. Obsch. Khim.*, 22 (1952) 361.
100 KAKIHANA, H. AND SILLEN, L. G., *Acta Chem. Scand.*, 10 (1956) 985.
101 DELIMARSKI, YU. K., SHEIKO, I. N. AND FESCHENKO, V. G., *Zhur. Fiz. Khim.*, 29 (1955) 1489.
102 MARKOV, B. F. AND DELIMARSKI, YU. K., *ibid.*, 31 (1957) 2589.
103 MARKOV, B. F., DELIMARSKI, YU. K. AND PANCHANKO, I. D., *J. Polym. Sci.*, 31 (1958) 263.
104 SHEIKO, I. N. AND DELIMARSKI, YU. K., *Ukrain. Khim. Zhur.*, 23 (1957) 713; *Chem. Abs.*, 52 (1958) 14384.
105 SMIRNOV, M. V. AND CHUKREEV, N. YA., *Zhur. Neorg. Khim.*, 4 (1959) 2536; *Zhur. Fiz. Khim.*, 32 (1958) 2165.

REFERENCES

106 FURLEY, E. AND WILKINSON, K. L., *J. Inorg. Nucl. Chem.*, 14 (1960) 123.
107 KELLS, M. C., HOLDEN, R. B. AND WHITMAN, C. I., *J. Amer. Chem. Soc.*, 79 (1957) 3925.
108 MESSERKNECHT, C. AND BILTZ, W., *Z. Anorg. Chem.*, 148 (1925) 157.
109 BERGSTROM, F. W., *J. Amer. Chem. Soc.*, 50 (1928) 657.
110 FRICKE, R. AND RUSCKHAUPTE, F., *Z. Anorg. Chem.*, 146 (1924) 103.
111 FRICKE, R. AND HAVESTADT, L., *ibid.*, 146 (1924) 121.
112 FRICKE, R. AND RODE, O., *ibid.*, 163 (1927) 31.
113 FRICKE, R. AND ROBKE, F., *ibid.*, 170 (1928) 25.
114 FRICKE, R., *ibid.*, 252 (1947) 173.
115 PRASED, S. AND SRIVOASTAVA, K. P., *J. Ind. Chem. Soc.*, 35 (1958) 793.
116 TUROVA, N. YA., NOVOSELOVA, A. V. AND SEMENENKO, K. N., *Zhur. Neorg. Khim.*, 4 (1959) 2204.
117 NESPITAL, W., *Z. Phys. Chem.*, 16 (1932) 153.
118 TUROVA, N. YA., NOVOSELOVA, A. V. AND SEMENENKO, K. H., *Zhur. Neorg. Khim.*, 5 (1960) 117.
119 SILBER, M. P., *Ann. Chim. France*, 7 (1952) 182.
120 ULICH, H. AND NESPITAL, W., *Z. Electrochem.*, 37 (1931) 559.
121 MILIOTIS, J. A., GALINOS, A. G. AND TSANGARIS, J. M., *Bull. Soc. Chim. France* (1961) 1413.
122 GROSSE, A. V. AND IPATIEFF, V. M., *J. Org. Chem.*, 1 (1937) 559.
123 BREDERECK, H., LEHMANN, G., SCHONFIELD, C. AND FRITZSCH, E., *Ber.*, 72 (1939) 1414.
124 COLCLOUGH, R. O., GEE, G., HIGGINSON, W. C. E., JACKSON, J. B. AND LITT, M., *J. Polym. Sci.*, 34 (1959) 171.
125 WOOD, G. B. AND BRENNER, A., *J. Electrochem. Soc.*, 104 (1957) 29.
126 MULLER, R., *Z. Anorg. Chem.*, 142 (1925) 131.
127 BILTZ, W. AND KEUNECKE, E., *ibid.*, 147 (1925) 124.
128 PAJEAU, R., *Compt. Rend.*, 204 (1937) 1347; *ibid.*, 207 (1938) 344 and 1420; *Bull. Soc. Chim. France* (1939) 1187.
129 TUROVA, N. YA., NOVOSELOVA, A. V. AND SEMENENKO, K. N., *Zhur. Neorg. Khim.*, 5 (1960) 94.
130 CHAIGNEAU, M., *Bull. Soc. Chim. France* (1957) 886.
131 CUPR, V. AND SALANSKY, H., *Z. Anorg. Chem.*, 176 (1928) 249.
132 SIDGWICK, N. V., *The Chemical Elements and Their Compounds*, Oxford University Press, London, 1950.
133 KOPELMAN, B. AND BENDER, H., *J. Electrochem. Soc.*, 98 (1951) 89.
134 SLOMAN, H. A., *J. Inst. Metals*, 49 (1932) 365.
135 METCALFE, W. J., PERRY, G. S. AND TURNER, K. H., *U.K.A.E.A. report*, AWRE 0–21/62 (1962).

CHAPTER 5

Complex Beryllium Compounds

Here we deal with the chemistry of complex beryllium compounds except for the complexes formed by beryllium halides which have already been covered (Chap. 4). First the factors which govern the stability of beryllium complexes will be considered. This will be followed by an account of the preparation and properties of the main types of beryllium complex, for instance, the derivitives of 1,3–diketones, the hydroxy acids, and the oxide carboxylates. Finally a short account of beryllium coordination polymers and the beryllium derivitive of phthalocyanine will be given.

1. Factors governing the stability of beryllium complexes

As in the complexes of the alkali and other alkaline earth metals, the bonding in beryllium complexes is essentially ionic. In such complexes the strength of binding should increase as the ionic charge of both metal ion and ligand increase, and also as the radius of the metal ion falls, giving an increased charge to radius ratio.

It has been shown by Born[1] for spherical gaseous ions that the energy of solution (*i.e.* of solvation) is expressed by:

$$E = \frac{e^2}{2r}(1 - \frac{1}{D}),$$

where D is the dielectric constant of the solvent and r the radius of the ion. Since for a reaction log K is related to energy, Born's equation suggests that the ratio of the square of the charge of an ion to its radius might be better for purposes of correlation than the simple charge to radius ratio[2], and there appears to be some experimental evidence for this[2]. Nevertheless, as our discussions are of a qualitative character, the simple charge to radius ratio will be used.

The objection to a simple correlation of complex stability with factors such as the charge to radius ratio of the coordinating metal ion is that the free energy of formation of the complex ($\triangle G$) which is directly related to the stability constant ($-\triangle G = RT\ln K$), is itself a sum of an enthalpy ($\triangle H$) and an entropy ($\triangle S$) term. Let us consider the chelation reaction:

$$M(H_2O)_a^{2+} + a/n\,L \rightleftharpoons ML_{a/n}^{2+} + aH_2O,$$

in which a water molecules are displaced from the metal M by a/n molecules of a chelating agent with n donor groups per molecule. Now the degree of hydration of the ion will be the higher the greater its charge to radius ratio (see discussion of the hydration of the Be^{2+} ion, p. 7), which in turn leads to an increase in the number of particles formed in the above reaction. As increasing the number of particles causes a favourable increase in the entropy change for the reaction ($\triangle G = \triangle H - T\triangle S$), an increase in the charge to radius ratio of the metal ion will lead to an increased stability constant for the complex, providing no marked change occurs in the enthalpy of reaction. Thus to state that a relationship exists between the stabilities of a series of complexes and the charge to radius ratio of the central metal atom, is to say that the stability of a complex depends mainly upon entropy factors.

There are major breakdowns in such simple generalisations concerning the relative stabilities of a series of complexes when appreciable changes occur in the enthalpies of reaction of different cations with a given ligand. Thus for the ethylenediaminetetra-acetic acid (E.D.T.A.) complexes of the alkaline earth metals, the thermodynamic data for which are given in Table 3[3], although the entropy change on complex formation increases as expected with the charge to radius ratio, the stability of the magnesium complex is lower than that of the calcium complex owing to the endothermic character of the Mg^{2+}–E.D.T.A. reaction. Apparently a strainless octahedron is most closely approached in the calcium complex and

TABLE 3

THERMODYNAMIC CONSTANTS FOR THE FORMATION OF THE E.D.T.A. COMPLEX OF Mg, Ca, Sr AND Ba.

	$\triangle G$	$\triangle H^*$	$\triangle S$
Mg	–11.65	3.64	50.5
Ca	–14.34	–6.45	26.9
Sr	–11.57	–4.11	25.4
Ba	–10.40	–4.83	19.0

that a smaller cation results in strains being set up which account for the increased $\triangle H$ value. Such strains would be expected to be even greater in the Be^{2+}–E.D.T.A. complex, the reaction to be more endothermic and the complex correspondingly less stable. In agreement with this, log K is 3.8 for the Be–E.D.T.A. complex[4] compared with 8.69, 10.96, 8.63 and 7.76 for the Mg–, Ca–, Sr– and Ba–E.D.T.A. complexes, respectively[5] (cf. ref. 5a).

$$(K = \frac{[MY^{2-}]}{[M^{2+}][Y^{4-}]}).$$

The stabilities of beryllium complexes relative to those of other bivalent metals has also been discussed by Fernelius and co-workers[6-12]. They find[6] stability in a series of acetylacetone complexes to be in the order Ba⟨Sr⟨Ca⟨Mg⟨Cd⟨Zn⟨Be⟨Hg, and suggest that their stability is mainly controlled by the ionic radius and the atomic number of the metal atom. Considering only the alkaline earths, the above order is that expected on the basis of ionic radii and holds for all 1,3–diketone complexes. However, the position of magnesium in this series varies (presumably beryllium behaves similarly but quantitative data are missing) for complexes with ligands such as malic, tartaric and nitrilotriacetic acids. The reason is probably variations in the enthalpy of complex formation similar to that found with the E.D.T.A. complexes of magnesium and beryllium.

In Table 4 are given thermodynamic data[7] for the formation of acetylacetone complexes of various bivalent metals. It should be

---

* $\triangle H$ values listed above were determined by direct calorimetry.

noted that the $\triangle H$ and $\triangle S$ values were estimated from the variation of the formation constant with temperature over a 30° temperature range, and are thus probably less accurate than the direct calorimetric values of Care and Staveley[3]. The high stability of the beryllium complex is mainly due to the large and favourable entropy term. In contrast to the E.D.T.A. complexes, there appear to be no steric difficulties in the formation of the acetylacetone complexes of magnesium and beryllium, both reactions being exothermic.

TABLE 4

THERMODYNAMIC QUANTITIES FOR THE SUCCESSIVE STAGES OF FORMATION OF M^{2+}-ACETYLACETONE COMPLEXES.

	$\triangle G_1$	$\triangle G_2$	$\triangle H_1$	$\triangle H_2$	$\triangle S_1$	$\triangle S_2$
Be^{2+}	−10.4	−9.0	−2.0	−6.9	29	5
Mg^{2+}	−5.0	−3.5	−1.8	−4.3	11	2.5
Ni^{2+}	−8.4	−6.3	−6.7	−6.3	12	0
Cu^{2+}	−11.0	−9.1	−4.7	−6.6	22	9

Note: for $Ni^{2+} \triangle G_3 = -3.3$, $\triangle H_3 = 6.7$ and $\triangle S_3 = -12$.

The actual formation constants for the beryllium acetylacetonate complex[6-8] are $\log K_1 = 7.8$, $\log K_2 = 6.69$, $\log \beta_2 = 14.5$ (30°) and $\log K_1 = 7.93$, $\log K_2 = 6.96$, $\log \beta_2 = 14.89$ (10°) at ionic strengths → 0. The stability of the complexes are increased by employing 50% dioxan[9] ($\log K_1 = 9.0$, $\log K_2 = 7.7$, $\log \beta_2 = 16.7$ or 75% dioxan[10] ($\log K_1 = 12.36$, $\log K_2 = 10.94$, $\log \beta_2 = 23.20$) as solvent. Formation constants for other beryllium complexes with ligands related to acetylacetone, e.g. dibenzoylmethane and various tropolone derivitives, have also been recorded by Fernelius et al.[9-12].

2. *Beryllium complexes with 1,3-diketones*

Beryllium acetylacetonate is the best known member of this class. First described by Combes in 1894[13], it is prepared by addition of an ammoniacal solution of acetylacetone to an aqueous solution of a beryllium salt; usually the precipitated acetylacetonate is dissolved in the minimum quantity of benzene and reprecipitated by the addition of petroleum ether[14]:

$$BeCl_2 + 2C_5H_8O_2 + 2NH_3 = Be(C_5H_7O_2)_2 + 2NH_4Cl$$

It is colourless, monoclinic and crystalline, m.p. 108.5°, b.p. 270°, soluble in organic solvents, insoluble in water, decomposed by hot water, acids and alkalies. It is given the cyclic structure:

this being in agreement with infrared spectroscopic studies[15].

Numerous other crystalline 1,3-diketone complexes of beryllium have been reported, e.g. those from benzoylacetone (m.p. 210–11°) and dibenzoylmethane (m.p. 214–15°)[16], m- and p-nitrobenzoylacetone (m.p. 207–8 and 243°)[17], heptanedione-2,4 (m.p. 18°), 3-acetylhexanone-2 (m.p. 135–37°) and 3-ethylpentanedione-2,4 (m.p. 140–1°)[18], 5,5-dimethylcyclohexane-1,3-dione (m.p. 315°)[19], acetyltrifluoroacetone (m.p. 112°) and benzotrifluoroacetone (m.p. 143–4°)[20]. β-Ketoesters such as ethylacetoacetate[16] form similar beryllium compounds to those formed by 1,3-diketones.

The tetrahedral arrangement of the bonds in 4-covalent beryllium compounds such as those formed with 1,3-diketones, was first demonstrated by the classical researches of Mills and Gotts[21]. Following the work of Lowry[22] on the mutarotation of the beryllium derivitive of benzoyl camphor, the former authors unambiguously resolved the beryllium derivitive of benzoylpyruvic acid by crystallization of the brucine salt under carefully controlled conditions. Although benzoylpyruvic acid is itself inactive, its beryllium derivitive should exist in optically active forms if the beryllium valencies have a tetrahedral configuration and if the carbonyl oxygen is directly united to the metal to form a spirocyclic compound.

[Structure: Be complex with two PhCOCHCO₂H-derived ligands]

The resolution of other beryllium salts has also been reported, *e.g.* the benzoylacetone complex[23, 24].

X-ray studies of beryllium acetylacetonate have shown that the oxygens are tetrahedrally arranged around the beryllium and that, within experimental error, the acetylacetonate radicle is planar[25]. The structure of 1,3-diketone complexes of beryllium can thus be considered as firmly proved.

3. Beryllium complexes with hydroxy acids

The best known examples of this group of complexes are those with aromatic *o*-hydroxy acids, in particular salicylic

[salicylic acid structure]   and 5-sulphosalicylic acid   [5-sulphosalicylic acid structure]

with which beryllium forms 1 : 1 and 1 : 2 metal : ligand complexes. The 1 : 1 salicylic acid complex has been obtained both in solution[26, 27] and in the solid state. The solid has been formulated[28] as a dihydrate $Be(C_6H_4O.CO_2).2H_2O$ in which the hydrogens on both the hydroxyl and carboxyl groups have been replaced by the beryllium. The nature of this compound appears to be in some doubt, however, other authors formulate it as a trihydrate, $Be(C_6H_4O.CO_2).3H_2O$[29], or as $BeOH(C_6H_4OH.CO_2).2H_2O$[30]. When this 1 : 1 complex is dissolved in water, it is considered to ionize into Be^{2+} cations and $Be(C_6H_4O.CO_2)_2^{2-}$ anions[28]:

$$2Be(C_6H_4O.CO_2) \rightleftharpoons Be^{2+} + Be(C_6H_4O.CO_2)_2^{2-}$$

References p. 79

The latter anion possesses a chelate structure[29, 31].

$$\left[H_4C_6 \underset{CO_2}{\overset{O}{\diagdown}} Be \underset{CO_2}{\overset{O}{\diagup}} C_6H_4 \right]^{2-}$$

Formation of 1 : 1 and 1 : 2 complexes in solution is well defined with 5-sulphosalicylic acid. Spectrophotometric studies[32, 33] show that the 1 : 1 complex is formed between pH 2–6, evidence also being given for the displacement of the phenolic hydrogen on complex formation:

$$Be^{2+} + HSSA^{2-} \rightleftharpoons BeSSA^- + H^+,$$

where $HSSA^{2-} = {}^-O_3.S.C_6H_3.OH.CO_2^-$

Above pH 7, the 1 : 1 complex is transformed into the 1 : 2 complex, half the beryllium being precipitated as beryllium hydroxide.

$$2BeSSA^- + 2OH^- \rightleftharpoons Be(SSA)_2^{4-} + Be(OH)_2$$

Evidence for the formation of 1 : 1 and 1 : 2 complexes have also been obtained with beryllium and *o*-cresotic[34, 35] and 1-hydroxy-2-naphthoic acids[36, 37]. The latter complexes are insoluble in water, the 1 : 1 complex ($BeC_4H_6O_3.2H_2O$) being precipitated at pH 5.5 and the 1 : 2 complexes, $H_2[Be(C_{11}H_6O_3)_2]$ and $K_2[Be(C_{11}H_6O_3)_2]$, between pH 8–9.

Examples of beryllium complexes with aliphatic hydroxy acids are those with citric acid. Three 1 : 1 complexes are formed in the range pH 3–4, BeH_2Cit^+, BeHCit and $BeCit^-$[38]. The charges on the complexes indicate that the hydrogen on the hydroxyl group is not displaced on complex formation. This is presumably due to the acidity of the OH groups in citric being lower than the phenolic OH groups in salicylic acid. There is also evidence that above pH 8 the complex obtained contains two beryllium atoms per citrate molecule[39].

Complexes with more than one beryllium atom per ligand molecule are also formed with tartaric acid in alkaline solution[40-43]. Although their exact nature is still unknown, they are sufficiently stable to account for the solubility of beryllium hydroxide in aqueous sodium tartrate.

4. Beryllium oxide carboxylates (basic carboxylates)

The oxide carboxylates, $Be_4O(R.CO_2)_6$, form one of the most important groups of compounds in beryllium chemistry, the best known member being the oxide acetate which was first made by Urbain and Lacombe[44]. There are four main methods by which compounds of this type may be prepared[45].

(a) The action of the organic acid (or acid anhydride[45a]) on beryllium oxide or hydroxide followed by evaporation to a solid or an oily liquid. The oxide carboxylate is extracted with chloroform and recrystallised from the solvent. This method does not give the oxide formate, only the normal formate being formed.

(b) The action of the organic acid on the oxide carbonate. This is a way of obtaining the oxide formate[46, 47]; it can also be made by heating[48] the normal formate to 250–260°.

(c) The action of the organic acid or acid anhydride on anhydrous beryllium chloride[49, 50]. This results in the oxide carboxylate except under very carefully controlled conditions when normal carboxylates are produced.

(d) The action of the organic acid or acid chloride on another beryllium oxide carboxylate[47]. This provides a useful method for the preparation of mixed oxide carboxylates. These mixed compounds are also formed by melting the individual carboxylates together; Hardt[51] has prepared the whole range of compounds $Be_4O(CH_3.CO_2)_x(CH_3.CH_2CO_2)_y$ (y + x = 6) by either melting together the oxide acetate and propionate, by action of acetyl chloride on the propionate or by action of propionic acid on the oxide acetate.

The oxide carboxylates are non-electrolytes, soluble in organic solvents such as chloroform or benzene, insoluble in cold water, possess sharp melting points and can usually be sublimed or distilled without decomposition. Thus the oxide acetate melts at 285–6°, boils at 330° and can be sublimed at its melting point in a vacuum. The oxide carboxylates are stable towards heat and oxidation except under drastic conditions, and are only slowly hydrolysed by hot water. Treatment with mineral acids gives the corresponding beryllium salt and free organic acid.

In Fig. 5 is shown the structure of beryllium oxide acetate as determined by X-ray analysis[52-54]. The molecule has tetrahedral symmetry, the central oxygen atom being surrounded tetrahedrally by four beryllium atoms and the six acetate groups are attached symmetrically to the six edges of the tetrahedron. The crystal symmetry of the oxide acetate is in accord with its crystallizing in the cubic system. Replacement of a hydrogen by a methyl will induce some dissymmetry, as is shown by oxide propionate and isobutyrate crystallizing in the monoclinic and triclinic systems,

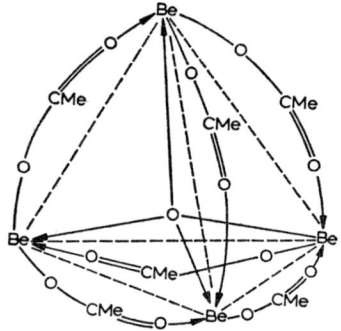

Fig. 5. Structure of beryllium oxide acetate.

respectively. This loss of symmetry is also shown by some of the heavier oxide carboxylates possessing lower melting points than the oxide acetate, for instance propionate 138°, isobutyrate 88–9°, n-butyrate 26°.

As might be expected from its structure, beryllium oxide acetate has a number of transitions in the solid phase. The stable modification is cubic, but at least two monoclinic forms are known. A high temperature monoclinic form is obtained[55-58] on heating the cubic modification to 145–150°; it is definitely monoclinic, not rhombohedral as originally suggested by Saito[55, 58]. A second monoclinic modification appears to result from sublimation[55, 58, 59] or recrystallisation of the oxide acetate from hot tetralin[60]. In addition to these phase changes an order-disorder transition of the cubic form takes place at 40°[61-63]. It does not involve a phase change as the crystals can be repeatedly heated and cooled through the transition temperature without any recrystallisation taking place[61]. The transition is probably due to a rotation of the carbonyl oxygens around the C–C bonds.

Beryllium oxide carboxylates can form clathrate compounds[64] with addenda such as sulphur dioxide[65, 66] or dioxan[67]. Semenenko[66] suggests that formation of such compounds is promoted by the coordination saturation of the oxide acetate molecule making dative bond formation difficult or impossible without an extensive breakdown of the bond system within the molecule. The high polarizabilities of the large oxide carboxylate molecules (51.6 cc for $Be_4O(CH_3CO_2)_6$ and 125 cc for $Be_4O(C_6H_5CO_2)_6$ allows the London forces between the oxide salt and the complexing component to be as large as in other clathrate compounds. Semenenko further suggests that the limiting structure of such clathrate complexes should be $Be_4O(CH_3CO_2)_6 \cdot 2X$, where X is an approximately spherical molecule of about 2Å radius. Such a limiting formula is suggested as arising from the fact that the surface of a Be_4O tetrahedron presents four projections and four depressions.

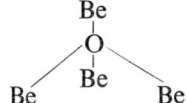

Two neighbouring oxide carboxylate molecules may be so packed into the crystal lattice that either the projections on one molecule fit into the depressions of the second or that the depressions come together to form a closed cavity. The maximum number of cavities is thus approximately twice the number of oxide acetate molecules present.

Ammonia and amines form a number of compounds with beryllium oxide acetate by direct interaction with liquid ammonia or amine. Liquid ammonia[65, 68] gives $Be_4O(CH_3CO_2)_6 \cdot 12NH_3$ at $-33°$, which readily loses ammonia to give $Be_4O(CH_3CO_2)_6 \cdot 4NH_3$ and $Be_4O(CH_3CO_2)_6 \cdot 3NH_3$. The last two compounds are completely decomposed at 180° into ammonia and the oxide acetate. They

References p. 79

dissolve with decomposition in water, and absolute alcohol; from solution in chloroform the oxide acetate is recoverable. Similar compounds have also been obtained with methylamine, ethylamine, butylamine and pyridine[65, 68-71]. All these are relatively unstable, decomposing into their components on heating or on dissolution in solvents such as chloroform.

Steric factors appear to play an important part in the formation of these compounds of amine and oxide acetate. *e.g.* monomethylamine will combine with beryllium oxide acetate but di- and tri-ethylamine will not. In this respect these complexes behave as if they were inclusion compounds, but their other properties make it doubtful whether they can be considered as clathrates. Thus molecules such as acetonitrile, ethyl bromide or nitromethane, whose geometrical configuration closely resembles that of ethylamine, do not form any complexes with beryllium oxide acetate. Moreover, the decomposition temperature of the amine compounds is independent of the boiling point of the amine, and the decomposition is accompanied by marked heat changes, both effects being untypical of true clathrate compounds. If Semenenko's suggestion[66] is accepted, then molecules such as pyridine or butylamine would be too big to fit into the holes in the oxide acetate lattice, so that genuine inclusion compounds could not be formed.

Infrared measurements[68] have been interpreted as indicating that beryllium–nitrogen bonds occur in these amine–oxide acetate compounds, indicating that some redistribution of bonds must have taken place within the oxide acetate molecule. It is possible that not all of the acetates are now acting as chelate groups and that some charged acetate ions are present. Indirect support for this suggestion is provided by the electrical conductivity of solutions of beryllium oxide acetate in liquid ammonia.

Similar amine complexes are formed with beryllium oxide propionate but the oxide formate, like the zinc oxide acetate, is decomposed by ammonia or amines[72].

The reaction of beryllium oxide acetate with alcohols affords a number of novel and as yet incompletely characterised materials. On boiling the oxide acetate with ethanol or methanol a clear solution is first formed which gives a voluminous white precipitate on standing, the mother liquor containing acetic acid and part of the beryllium. These results have been interpreted[73, 74] as indicating the reaction:

$$Be_4O(CH_3CO_2)_6 + 2ROH =$$
$$3Be(OR)CO_2CH_3 + 2CH_3CO_2H + Be(OH)CO_2CH_3,$$

the last compound being responsible for the beryllium which remains in solution. $Be(OR)CO_2CH_3$ is insoluble in alcohol, and is hydrolysed by water or in moist air with loss of alcohol.

In addition to the above main reaction two secondary reactions also occur, the acetic acid formed above being esterified:

$$CH_3CO_2H + ROH \rightleftharpoons CH_3CO_2R + H_2O$$

and the liberated water then reacting with the beryllium compound precipitated in the first reaction:

$$Be(OR)CO_2CH_3 + H_2O = Be(OH)CO_2CH_3 + ROH$$

Reaction can continue past the point indicated by these equations. Thus from beryllium oxide acetate and n-butanol[74] the compound $Be(OC_4H_9)CO_2CH_3$ first separates but dissolves again on heating. On cooling the solution less of the compound is precipitated, and, after 10–12 cycles, precipitation on cooling ceases. The viscosity of the solution increases rapidly on boiling; it reaches a maximum after about 72 h when an amorphous precipitate begins to form. This has a composition approximating to $3Be(OH)CO_2CH_3 \cdot Be(OC_4H_9)CO_2CH_3$, probably containing

$$-\overset{|}{Be}-O-\overset{|}{C}=O\rightarrow\overset{|}{Be}-O-\overset{|}{C}=O\rightarrow\overset{|}{Be}- \text{ units.}$$

Similar reactions have been observed by Hardt[51, 75], although his conclusions are somewhat different. He suggests that the compound precipitated from alcoholic oxide acetate are alcohol adducts of higher oxide acetates, $[Be_4O(CH_2CH_3)_{8-2m}]n$, where $m \rangle 1$, $n \geqslant 2.4$. Although it is not possible to decide between these conflicting

References p. 79

views, clearly there is agreement that beryllium oxide acetate gives polymeric compounds in alcoholic systems.

It is of interest that the compound $Be_4O_3(CH_3CO_2)_2$ has been claimed as being formed by the slow hydrolysis of the oxide acetate at 20°[76]. This material may well be similar to the compounds discussed in the previous paragraphs. The potentiometric measurements of Prytz[77] also seem to indicate that there are some unusual features connected with aqueous beryllium acetate solutions.

5. *Beryllium co-ordination polymers*

The first attempt to prepare a beryllium containing co-ordination polymer was by Wilkins and Wittbecker[78] by the reaction of bis-(1,3-diketones) with beryllium and other 4-co-ordinate metals. They were unable to obtain chains of sufficient length to give any plasticity to the polymer, a difficulty encountered by later workers. Work along similar lines has recently been undertaken by Kluiber and Lewis[79]. By the reaction of bis (1,3-dicarboxyl) compounds with beryllium salts in a neutral medium, macrocyclic compounds (*e.g.* I and II) are obtained. When these are heated above their melting points they polymerize to give X-ray amorphous material (III).

An example is a polymer in which $R' = CH_3$ and $R = (CH_2)_8$, prepared by heating (I) to 200° at atmospheric pressure for 15 min. This material is flexible at room temperature and has a glass transition temperature of 35°.

Polymers of the above type are only known with tetrafunctional ligands, no macrocycles involving beryllium and a difunctional ligand have yet been obtained. In general, the initial formation of a macrocycle of type (I) is favoured when R is six or more carbons long, shorter chains favour formation of type (II) macrocycles.

Another method for preparing beryllium polymers is to substitute two of the acetate groups in beryllium oxide acetate with a dibasic acid[80]:

$$n Be_4O(CH_3CO_2)_6 \ + \ n X \cdot OC \cdot R' \cdot COX =$$

$$= \left[\begin{pmatrix} Be_4O \\ (CH_3CO_2)_4 \end{pmatrix} \begin{pmatrix} -O \\ -O \end{pmatrix} C \cdot R' \cdot C \begin{pmatrix} O- \\ O- \end{pmatrix} \right]_n + \ 2n R \cdot CO \cdot X$$

i.e. one dibasic acid unit links two $Be_4O(CH_3CO_2)_4$ units together. The experimental procedure was to condense equimolecular quantities of the oxide acetate and the dibasic acid chloride in a hydrocarbon solvent such as benzene or toluene, and then coagulate the polymer by addition of a low boiling petroleum ether. Polymers containing only aliphatic acids showed a limited elasticity when plasticised with a little solvent and could be drawn out into short fibres. These linear polymers disproportionate slowly at room temperature, more rapidly on heating, to give a cross-linked polymer and monomeric oxide carboxylate.

$$\left[\begin{pmatrix} Be_4O \\ (CH_3CO_2)_4 \end{pmatrix} \begin{pmatrix} -O \\ -O \end{pmatrix} C \cdot R' \cdot C \begin{pmatrix} O- \\ O- \end{pmatrix} \right]_n =$$

$$= Be_4O(CH_3CO_2)_6 \ + \ \left[(Be_4O) \begin{pmatrix} -O \\ -O \end{pmatrix} C \cdot R' \cdot C \begin{pmatrix} O- \\ O- \end{pmatrix}_3 \right]_n$$

References p. 79

This reaction destroys any of the desirable plasticity possessed by the original linear polymer. If desired, the cross-linked polymers can be directly prepared by carrying out the initial condensation in a concentrated solution or by employing a higher reaction temperature.

A polymeric beryllium containing material has also been obtained by the interaction of anhydrous beryllium chloride with a polymeric silylamine $(Me_2SiNH.CH_2.CH_2.NH)_n$ in xylene[81], about one-third of the nitrogens being co-ordinated to the beryllium. The resulting resinous beryllium containing polymer is more stable towards hydrolysis than the original liquid polymeric silylamine.

6. Beryllium–phthalocyanine complexes

These complexes represent one of the few instances in which beryllium forms a stable complex containing only beryllium–nitrogen bonds. Like other metal–phthalocyanine complexes the beryllium compound is obtained by directly dissolving metallic beryllium in phthalonitrile[82]. This reaction yields the anhydrous 1 : 1 complex (PcBe), which forms a dihydrate on standing in moist air. PcBe is slightly soluble in solvents such as quinoline or pyridine, solvated crystals being obtained on recrystallization. Concentrated sulphuric acid dissolves PcBe with decomposition, free phthalocyanine being deposited on diluting the solution with water.

X-ray analysis has shown PcBe to possess a similar planar structure to other metal phthalocyanines[83].

The geometry of the molecule forces the beryllium to adopt a planar configuration in contrast to its normal tetrahedral arrangement. That such a configuration is unstable is shown by the fact that PcBe forms a dihydrate when presumably two of the nitrogen–beryllium bonds in the phthalocyanine are broken and replaced by bonds between beryllium and the water molecules. Such an arrangement allows the beryllium to adopt a tetrahedral configuration. It is significant that PcMg is the only other phthalocyanine complex to form a hydrate.

REFERENCES

1 BORN, M., Z. Physik, 1 (1920) 45.
2 MARTELL, A. E. AND CALVIN, M., Chemistry of the Metal Chelate Compounds, Prentice Hall, New York, 1952.
3 CARE, R. A. AND STAVELEY, L. A. K., J. Chem. Soc., (1956) 4751.
4 FAIRHALL, A. W., Radiochemistry of Beryllium, U.S.A.E.C. report NAS-NS3013 (1960).
5 SCHWARZENBACH, G., GUT, R. AND ANDEREGG, G., Helv. Chim. Acta, 37 (1954) 937.
5a BAES, C. F., J. Inorg. Nucl. Chem., 24 (1962) 707.
6 IZATT, R. M., FERNELIUS, W. C. AND BLOCK, B. P., J. Phys. Chem., 59 (1955) 80.
7 Idem, ibid., 59 (1955) 235.
8 Idem, ibid., 58 (1954) 1133.
9 BRYANT, B. E. AND FERNELIUS, W. C., J. Amer. Chem. Soc., 76 (1954) 1696.
10 VAN UITERT, L. G., FERNELIUS, W. C. AND DOUGLAS, B. E., ibid., 75 (1953) 2736.
11 BRYANT, B. E. AND FERNELIUS, W. C., ibid., 76 (1954) 3783.
12 BRYANT, B. E. AND FERNELIUS, W. C. AND DOUGLAS, B. E., ibid., 75 (1953) 3784.
13 COMBES, A., Compt. Rend., 119 (1894) 1222.
14 ARCH, A. AND YOUNG, R. C., Inorganic Syntheses. Vol. 2, McGraw-Hill, New York, 1946.
15 DUVAL, C., FREYMAN, R. AND LECOMPTE, J., Bull. Soc. Chim. France, (1952) 106.
16 BOOTH, H. S. AND PIERCE, D. G., J. Phys. Chem., 37 (1933) 69.
17 BURGESS, H., J. Chem. Soc., (1927) 2017.
18 PRZHEVAL'SHII, E. S. AND MORSEEVA, L. M., Vestnik Moskov. Univ., 14 (1959) 203; U.K.A.E.A. report AWRE/Trans/10, 1960.
19 WEYGAND, C. AND FORKEL, H., Ber., 59 (1926) 2243.
20 BERG, E. W. AND TRUMPER, J. T., J. Phys. Chem., 64 (1960) 487.

21 MILLS, W. H. AND GOTTS, R. A., *J. Chem. Soc.*, (1926) 3121.
22 BURGESS, H. AND LOWRY, T. M., *ibid.*, (1924) 2081; LOWRY, T. AND TRAILL, R. C., *Proc. Roy. Soc.*, 132 (1931) 398.
23 BUSCH D. H. AND BAILAR, J. C., *J. Amer. Chem. Soc.*, 76 (1954) 5352.
24 FERRONI, E. AND CINI, R., *ibid.*, 82 (1960) 2477.
25 AMIRTHALINGAN, V., PADMANABHAN, V. M. AND SHANKER, J., *Acta Cryst.*, 13 (1960) 201; see also BULLEN, G. J., *ibid.*, 10 (1957) 143.
26 SCHUBERT, J. AND LINDENBAUM, H., *J. Biol. Chem.*, 208 (1954) 359.
27 VARMA, I. D. AND MEKROTRA, R. C., *J. Ind. Chem. Soc.*, 35 (1958) 381.
28 JONES, F. E., HAMER, W. E., DAVIES, C. W. AND BURY, C. R., *J. Phys. Chem.*, 34 (1930) 563.
29 ASMUSSEN, R. W. AND RANKE MADSEN, E., *Z. Anorg. Chem.*, 212 (1933) 321.
30 VENKATASUBRAMANIAN, K., *Anal. Chem.*, 32 (1960) 1052.
31 ROSENHEIM, A. AND LEHMANN, F., *Ann.*, 440 (1924) 153.
32 MEEK, H. V. AND BANKS, C. V., *J. Amer. Chem. Soc.*, 73 (1951) 4108.
33 BANKS, C. V. AND SINGH, R. S., *ibid.*, 81 (1959) 6159.
34 TRIPATHI, S. C., *J. Prakt. Chem.*, 12 (1961) 215.
35 BHATTACHARYA, A. K. AND RAO, G. S., *J. Sci. Ind. Res. India*, 18B (1959) 351.
36 Idem, *ibid.*, 18B (1959) 476.
37 AGARUAL, R. P. AND MEHROTRA, R. C., *J. Less Common Metals*, 3 (1961) 398.
38 FELDMAN, I. TORIBARA, T. Y., HAVILL, J. R. AND NEUMAN, W., *J. Amer. Chem. Soc.*, 77 (1955) 878.
39 FELDMAN, I., NEUMAN, W. F., DANLEY, R. A. AND HAVILL, J. R., *ibid.*, 73 (1951) 4775.
40 BRITTON, H. T. S., *J. Chem. Soc.*, (1926) 291.
41 DELSAL, J. L., *Compt. Rend.*, 198 (1934) 2076.
42 PEYCHES, I., *ibid.*, 198 (1934) 1778.
43 ZOLOTUKHIN, V. K., *Zhur. Neorg. Khim.*, 1 (1956) 2676.
44 URBAIN, G. AND LACOMBE, H., *Compt. Rend.*, 133 (1901) 874.
45 MOELLER, T., *Inorganic Syntheses*, 3 (1950) 4. McGraw-Hill, New York.
45a KAWECKI, H. C., *USP* 2,641,611.
46 LACOMBE, H., *Compt. Rend.*, 134 (1902) 772.
47 TANATER, S. AND KUROWSKI, E., *J. Russ. Phys. Chem. Soc.*, 39 (1907) 937.
48 NOVOSELOVA, A. V., SEMENENKO, K. N., KRASOVSKAYA, N. M. AND SIMANOV, YU. P., *Vestnik Moskov. Univ.*, 11 (1956) 86; *Chem. Abs.*, 50 (1956) 15320.
49 FIELD, G. W., *J. Amer. Chem. Soc.*, 61 (1939) 1817.
50 FUNK, H. AND ROMER, F., *Z. Anorg. Chem.*, 239 (1938) 288.
51 HARDT, H. D., *Z. Anorg. Chem.*, 314 (1962) 210.
52 BRAGG, W. H. AND MORGAN, G. T., *Proc. Roy. Soc.*, 104 (1923) 437.
53 MORGAN, G. T. AND ASTBURY, W. T., *ibid.*, 112 (1926) 441.
54 PAULING, L. AND SHERMAN, J., *Proc. Natl. Acad. Sci.*, 20 (1934) 340.
55 WATANABE, T. AND SAITO, Y., *Nature*, 163 (1949) 225.
56 HARDT, H. D. AND HENDUS, H., *Z. Anorg. Chem.*, 270 (1952) 298.

REFERENCES

57 JAFFRAY, J., *Compt. Rend.*, 234 (1952) 1539.
58 SAITO, Y., *X-Sen.*, 7 (1952) 9; *Chem. Abs.*, 46 (1952) 10760.
59 SEMENENKO, K. M. AND GORDEEV, I. V., *Zhur. Neorg. Khim.*, 4 (1959) 952.
60 HENDUS, H. AND HARDT, H. D., *Z. Anorg. Chem.*, 286 (1956) 265.
61 PRESTON, G. D. AND TROTTER, J., *Nature*, 151 (1943) 166.
62 WATANABE, T., SAITO, Y. AND KOYAMA, H., *ibid.*, 164 (1949) 1040.
63 JAFFRAY, J., *Compt. Rend.*, 225 (1947) 106; 226 (1948) 397.
64 POWELL, H. M., *Rec. Trav. Chim.*, 75 (1956) 885.
65 GRIGOR'EV, A. I., NOVOSELOVA, A. V. AND SEMENENKO, K. N., *Zhur. Neorg. Khim.*, 2 (1957) 1374.
66 SEMENENKO, K. N., *ibid.*, 5 (1960) 2687.
67 NOVOSELOVA, A. V., SIMANOV, YU. P., SEMENENKO, K. N. AND KRASOVSKAYA, N. N., *ibid.*, 1 (1956) 2280.
68 GRIGOR'EV, A. I. AND NOVOSELOVA, A. V., *ibid.*, 5 (1960) 2280.
69 GRIGOR'EV, A. I. AND SEMENENKO, K. N., *ibid.*, 3 (1958) 2806.
70 GRIGOR'EV, A. I., NOVOSELOVA, A. V. AND SEMENENKO, K. N., *ibid.*, 3 (1958) 1599.
71 NOVOSELOVA, A. V., SIMANOV, YU. P., SEMENENKO, K. N. AND KRASOVSKAYA, N. N., *ibid.*, 1 (1956) 690.
72 GRIGOR'EV, A. I. AND NOVOSELOVA, A. V., *ibid.*, 4 (1959) 2640.
73 NOVOSELOVA, A. V., SIMANOV, YU. P., SEMENENKO, K. N. AND KRASOVSKAYA, N. N., *ibid.*, 1 (1956) 196.
74 GRIGOR'EV, A. I., NOVOSELOVA, A. V. AND SEMENENKO, K. N., *ibid.*, 2 (1957) 2067.
75 HARDT, H. D., *Z. Anorg. Chem.*, 286 (1956) 254; 292 (1957) 53; 293 (1957) 46.
76 QUINET, M. L., *Compt. Rend.*, 218 (1944) 675.
77 PRYTZ, M., *Arch. Math. Naturvidenskab*, 45 (1942) 155.
78 WILKINS, J. P. AND WITTBECKER, E. L., *USP* 2,659,711 (1953).
79 KLUIBER, R. W. AND LEWIS, J. W., *J. Amer. Chem. Soc.*, 82 (1960) 5777.
80 MARVEL, C. S. AND MARTIN, M. M., *ibid.*, 80 (1958) 233.
81 MINNE, R. AND ROCHOW, E. G., *ibid.*, 82 (1960) 5628.
81a BLOCK, B. P. *et al.*, *Inorg. Chem.*, 1 (1962) 860.
82 BARRETT, P. A., DENT, C. E. AND LINSTEAD, R. P., *J. Chem. Soc.*, (1936) 1719.
83 LINSTEAD, R. P. AND ROBERTSON, J. M., *ibid.*, (1936) 1736.

CHAPTER 6

Simple Binary Compounds of Beryllium

The compounds dealt with here are those with hydrogen, boron, carbon, nitrogen, oxygen and sulphur, together with some closely related derivitives of these. Intermetallic compounds of beryllium will not be considered as they fall outside the arranged scope of this book.

1. Beryllium hydride

Beryllium hydride, BeH_2, is prepared by action of ethereal lithium aluminium hydride on beryllium chloride or dimethylberyllium[1,2]. A white solid is obtained from which the complete removal of ether does not appear to be possible; the best material contains only 89% BeH_2[2,3]. A purer product (96%) is obtained by the pyrolysis of di–*tert.*–butylberyllium at 210°[3]:

$$(Me_3C)_2Be = BeH_2 + 2Me_2C{=}CH_2$$

Beryllium hydride made by pyrolysis is thermally relatively stable, and does not decompose below 240°. Indeed, its thermal stability is only slightly less than that of the MgH_2 produced by pyrolysing primary magnesium alkyls[4]. The hydride prepared by Schlesinger's lithium aluminium hydride method is less stable, its decomposition temperature is reduced to 125°, possibly owing to the ether present.

Beryllium hydride is a three dimensional polymer, considerably ionic in character, and containing formally Be^{2+} and H^- ions[5].

$$-H^-\cdots \underset{\underset{H^-}{|}}{\overset{\overset{H^-}{|}}{Be^{2+}}} \cdots H^- - \underset{\underset{H^-}{|}}{\overset{\overset{H^-}{|}}{Be^{2+}}} - H^- \cdots \underset{\underset{H^-}{|}}{\overset{\overset{H^-}{|}}{Be^{2+}}} \cdots H^- -$$

The material produced by Schlesinger's method is quantitatively decomposed by water:

$$BeH_2 + 2H_2O = Be(OH)_2 + H_2$$

But the pyrolytic material requires dilute acid for its complete hydrolysis.

Closely related to beryllium hydride is beryllium borohydride, $Be(BH_4)_2$, first prepared by Berg and Schlesinger[6] by action of diborane on dimethylberyllium:

$$3Be(CH_3)_2 + 4B_2H_6 = 3Be(BH_4)_2 + 2B(CH_3)_3$$

Alternatively, $Be(BH_4)_2$ can be obtained by action of lithium borohydride on beryllium chloride[7]; the reaction is carried out at 90–140° and the sublimate of $Be(BH_4)_2$ is trapped at —80°:

$$2LiBH_4 + BeCl_2 = Be(BH_4)_2 + 2LiCl$$

Beryllium borohydride melts at 123° and sublimes at 91°. Like $Al(BH_4)_3$, it inflames spontaneously in air and reacts vigorously with water:

$$Be(BH_4)_2 + 8H_2O = Be(OH)_2 + 2B(OH)_3 + 8H_2$$

With acids some diborane is produced:

$$Be(BH_4)_2 + 2HCl = BeCl_2 + H_2 + B_2H_6$$

The borohydride has an unsymmetrical bridge structure, in which

the Be–H bond is longer than the B–H bond[8]. It can be considered as an ionic compound, $Be^{2+}(BH_4^-)_2$, which has a high degree of covalent character owing to the deformation of the BH_4^- ions by the strongly polarising Be^{2+} ions. If beryllium hydride and borohydride are not formulated ionically then they must be formulated as electron deficient molecules, similar to the beryllium alkyls[9]. The latter, along with the organoderivitives of beryllium hydride are dealt with in Chapter 7.

References p. 89

2. Beryllium borides

As with many other metallic borides, it is not possible to account for the formulae of the beryllium borides in terms of the ordinary concepts of valency. Probably the boron atoms are linked together in chains, layers or three dimensional networks extending throughout the whole crystal. The beryllium atoms are then accommodated in the boron network. The related compound, CaB_6, possesses a sodium chloride structure[10] in which sodium ions are replaced by calcium and chloride ions by regular octahedra of boron atoms. Every octahedron is linked to six neighbouring octahedra at a distance of 1.72Å; this is also the interatomic distance within each octahedron. For a theoretical discussion of the bonding in this compound see ref. 11.

Three beryllium borides are known, Be_2B, BeB_2 and BeB_6, with BeB_4 as an additional possible phase[12-15a]. All are formed by direct interaction of the elements at 1400°. Be_2B is the most reactive of these borides, being readily hydrolysed by dilute acids to give boron hydrides. In colour it is copper-red and possesses a cubic, fluorite structure. The other beryllium borides are insoluble in acids.

These compounds have possible applications as refractory materials. Thus they are only slowly attacked on heating in oxygen or nitrogen due to the formation of protective coatings. Attack by carbon is more rapid because the beryllium carbide formed on the surface is non-protective.

3. Beryllium carbide

Be_2C is obtained either by direct interaction of the elements or by the reduction of beryllium oxide with carbon at temperatures above 1500°[16, 17]. It is a translucent crystalline material, varying in colour from amber to dark brown when traces of carbon are present, and resembling carborundum in its abrasive properties. It is hydrolysed by water or dilute acids to give methane, although it is not attacked by moist air if in massive form. In the absence of moisture and oxygen it can act as a refractory material, not being decomposed below 2000–2100°, when beryllium vapour and graphite are formed. Metallic coatings or silicious glazes can be employed to give beryllium carbide resistance to atmospheric attack at high temperatures.

Beryllium carbide has an antifluorite structure[18] and can be considered as a salt-like carbide containing Be^{2+} and C^{4-} ions, formally it is a simple substitutional derivative of methane. It should be noted that metallic methanides are only formed by small ions like Be^{2+}, because of the limited room available between the close packed C^{4-} ions. Thus neither Mg^{2+} or Ca^{2+} are small enough for the elements to form methanides and both magnesium and calcium form only acetylene derivatives with carbon *viz.* MgC_2, Mg_2C_3 and CaC_2.

4. Beryllium nitride

Be_3N_2 is prepared by the action of nitrogen or ammonia on beryllium metal or carbide at temperatures $\geqslant 1000°$[13, 16]. It is a white crystalline material, subliming in a vacuum at 2000° and melting at 2200° with dissociation. The stable modification has a cubic (anti-Mn_2O_3) structure[17]; a high temperature hexagonal modification is reported to be formed either by subliming the cubic form or by heating it to 1400°, the latter change being catalysed by traces of silicon compounds[16, 18, 19].

Beryllium nitride is decomposed by water, and more rapidly by acids or alkalis, with evolution of ammonia. Like the carbide, the nitride is susceptible to attack by water vapour or oxygen and it must be protected from these when employed as a high temperature refractory. On heating Be_3N_2 and Si_3N_4 to 1650–1850° in a stream of ammonia[16, 18], the compound $BeSiN_2$ is obtained, which resembles aluminium nitride in possessing the wurtzite structure.

References p. 89

5. Beryllium oxide

Beryllium oxide is prepared by the calcination of beryllium hydroxide or oxide carbonate. However, for certain applications such as sintering studies, spectroscopic standards or special nuclear applications, beryllium oxide of exceptionally high purity is required. This has been obtained by recrystallising beryllium oxalate and decomposing it thermally at 700–800°[23, 24]. An alternative starting material is beryllium oxide acetate which has been carefully purified, either by distillation or vacuum sublimation, or by extraction into chloroform, followed by its recrystallisation from this solvent. The oxide acetate is converted to sulphate which is then ignited to oxide at 1000°[23], or an aqueous solution of the partly hydrolysed acetate can be further hydrolysed to beryllium hydroxide[25]. This is best undertaken by slowly distilling off acetic acid from the solution when beryllium hydroxide is precipitated as spherulites of high density; their size is controlled by the time that they are kept in the precipitation zone. By controlled calcination (final temperature 960°) spherical particles of beryllium oxide are obtained, composed of small crystals 0.1 microns or less, which are classified in a narrow particle range. Such a material is well suited to the production of high density beryllia refractories. Calcination of the hydroxide above 960° results in a marked increase in the size of the beryllium oxide crystallites.

Beryllium oxide acetate may be burned in oxidising gases directly to beryllia. Although this procedure is conducive to production of uniform spherical particles it has not been developed because of the difficulties associated with collection[25].

An interesting procedure for preparation of pure beryllium oxide is to spray a molten salt mixture of 60 mole% LiF and 40 mole% BeF_2 at 800° with helium saturated at room temperature with water[24]. The beryllium fluoride is hydrolysed to oxide, which is recovered by dissolving the lithium fluoride in hot aqueous aluminium nitrate. Spectrographic analysis shows that, except for slight contamination due to incomplete removal of solvent and traces of nickel from the container, this procedure gives a product, the purity of which compares favourably with oxide produced by other methods. Single crystals of beryllium oxide can be grown from Li_2MoO_4 melts[26] or from $Be(OH)_2$ vapour[27], the latter produced by the reaction between beryllium oxide and water to give gaseous $Be(OH)_2$ (see p. 13). The oxide crystallizes with a wurtzite structure in which the oxygen atoms are tetrahedrally arranged around the beryllium atoms. It thus differs from the other alkaline earth oxides which all have the sodium chloride structure. That beryllium oxide has the normally covalent wurtzite structure does not necessarily indicate that the beryllium–oxygen bonds are covalent in character. Indeed, it would be difficult to accommodate six oxygens around either a beryllium atom or a Be^{2+} ion. It is sufficient for most purposes to consider beryllium oxide as containing Be^{2+} and O^{2-} ions, but with the Be–O bonds possessing some covalent character due to the high polarizing power of the Be^{2+} ion and the high polarizability of the O^{2-} ions. A more sophisticated interpretation is that the bonding is $\sigma^2\pi^2$, with a double bond similar to that in ethylene, except for the polarity induced by the difference in electronegativity of the two atoms[28]. For further discussion of this problem see ref. 29.

Beryllium oxide melts at 2520–2570°; the extrapolated boiling point at 760 mm is 3900–4260°[30]. It possesses a high thermal conductivity which, in conjunction with its low thermal expansion, renders it a very thermostable material. In absence of water vapour, it is one of the least volatile of the refractory oxides, and its volatility is further reduced in presence of other non-volatile oxides, such as MgO, CaO, Al_2O_3, SiO_2, because of the formation of isomorphous mixtures or chemical compounds. Beryllium oxide is one of the most difficult oxides to reduce with carbon; this reaction does not start until ca. 1700°. As it also resists the attack of many reagents, at least at room temperature, and as it can be fabricated into dense shapes[16], beryllium oxide has many applications, both real and potential, as a refractory material.

Beryllium oxide forms binary compounds with a number of R_2O_3 oxides[31] and eutectic mixtures with many other oxides; for instance, beryllium and calcium oxides form a simple eutectic, m.p. 1405°, at 55 mole% BeO^{32}. Beryllium oxide is also a constituent of a number of glass forming mixtures (*e.g.* ref. 33).

Knudsen cell and mass spectrographic measurements[34] have shown that beryllium and oxygen atoms, and the polymeric species $(BeO)_3$ and $(BeO)_4$ are the principal constituents of beryllium oxide vapour between 1630–2130°. It is considered that these polymers probably possess ring structures.

6. Beryllium sulphide

This compound is prepared by the direct interaction of beryllium and sulphur at 1350°[35], or by action of hydrogen sulphide on beryllium metal at 900°. It has the zinc blende structure[36]; it is stable towards cold water, but is hydrolysed to beryllium oxide by boiling water. When heated in air, it is oxidised to the oxide and sulphur dioxide. An interesting property of the sulphide is that it shows fluorescence when it carries traces of impurity[37].

7. Beryllium selenide and telluride

These two compounds are formed by reaction between the elements. They resemble the sulphide in possessing the zinc blende structure[36].

8. The beryllides

Although alloys and intermetallic compounds of beryllium are not covered in this book, it is worth noting briefly that compounds of beryllium with high-melting transition elements such as tantalum, niobium, tungsten, molybdenum and zirconium (*e.g.* $TaBe_{12}$, Ta_2Be_{17}, $ZrBe_{13}$) have desirable properties as high-temperature materials[38]. Thus they are both resistant to oxidation and retain their mechanical strength up to *ca.* 1650°. They also possess a high specific heat and a good thermal couductivity.

SIMPLE BINARY COMPOUNDS OF BERYLLIUM

REFERENCES

1 SCHLESINGER, H. I., et al., *J. Amer. Chem. Soc.*, 73 (1951) 4585.
2 HEAD, E. L., HOLLEY, C. E. AND RABIDEAU, S. W., *ibid.*, 79 (1957) 3687.
3 COATES, G. E., AND GLOCKLING, F., *J. Chem. Soc.*, (1954) 2526.
4 WIBERG, E. AND BAUER, R., *Ber.*, 85 (1952) 593.
5 WIBERG, E., *Angew. Chem.*, 65 (1953) 16.
6 BERG, A. B. AND SCHLESINGER, H. I., *J. Amer. Chem. Soc.*, 62 (1940) 3425.
7 SCHLESINGER, H. I., BROWN, H. C. AND HYDE, E. K., *ibid.*, 75 (1953) 209.
8 BAUER, S. H., *ibid.*, 72 (1950) 622.
9 LONGUET-HIGGINS, H. C., *Quart. Rev.*, 11 (1957) 121; *J. Chem. Soc.*, (1946) 142.
10 PAULING, L. AND WEINBAUM, S., *Z. Krist.*, 87 (1934) 181.
11 LONGUET-HIGGINS, H. C. AND ROBERTS, M., *Proc. Roy. Soc.*, A224 (1954) 336; JOHNSON, R. W. AND DAANE, A. H., *J. Chem. Phys.*, 38 (1963) 425.
12 MARKOVSKII, L. Y., KONDRASHEV, Y. D. AND GORYACHEVA, I. A., *Dokl. Akad. Nauk, S.S.S.R.*, 101 (1955) 97.
13 MARKOVSKII, L. Y., KONDRASHEV, Y. D. AND KAPUTOVSKAYA, G. V., *Zhur. Obshch. Khim.*, 25 (1955) 1045.
14 SAMONOV, G. V. AND SEREBRYAKOVA, T. I., *Zhur. Priklad. Khim.*, 33 (1960) 563; see also SANDS, D. E., et. al., *Acta Cryst.*, 14 (1961) 309.
15 MARKEVICH, G. S. AND MARKOVSKII, L. Y., *ibid.*, 33 (1960) 1008 and 1667.
15a BECHER, H. J. AND SCHAFER, A., *Z. Anorg. Chem.*, 318 (1962) 304.
16 BEAVER, W. C., in *The Metal Beryllium*, p. 573. Ed. by White, D. W. and Burke, J. E. The American Society for Metals, Cleveland, 1955.
17 MALLETT, M. W., et al., *J. Electrochem. Soc.*, 101 (1954) 298.
18 STARITZKY, E., *Anal. Chem.*, 28 (1956) 915.
19 RABENAU, A. AND ECKERLIN, P., in *Special Ceramics*, p. 136. Ed. by Popper, P., Heywood and Co., London, 1960.
20 PAULUS, R., *Z. Phys. Chem.*, 22B (1933) 305.
21 ECKERLIN, P. AND RABENAU, A., *Z. Anorg. Chem.*, 304 (1960) 218.
22 CHIOTTI, P., *J. Amer. Ceram. Soc.*, 35 (1952) 123.
23 MOORE, R. E., *U.S.A.E.C. report ORNL-2938* (1960).
24 HARNS, W. O., in the *Proceedings of the beryllium oxide meeting, Oak Ridge National Laboratory*, 1960. *U.S.A.E.C. report TID-7602*.
25 KIRKPATRICK, W. J., ANDERSON, G. R. AND FUNSTON, E. S., *U.S.A.E.C. report APEX-684* (1961); AUSTERMAN, S. B., *J. Amer. Ceram, Soc.*, 46 (1963) 6.
26 AUSTERMAN, S. B. AND HOPKINS, A. R., *U.S.A.E.C. report NAA-SR-6425* (1962).
27 Idem, *U.S.A.E.C. report NAA-SR-6420* (1961); BUDNIKOV, P. P. AND SHISHKOV, N. V., *Dokl. Akad. Nauk S.S.S.R.*, 138 (1961) 1093.
28 COULSON, C. A., *Valence*, Oxford University Press, 1953, p. 317.
29 JEFFREY, G. A., PARRY, G. S. AND MOZZI, R. L., *J. Chem. Phys.*, 25 (1956) 1024; KEFFER, F. AND PORTIS, A. M., *ibid.*, 27 (1957) 675; O'SULLIVAN, W., *ibid.*, 30 (1959) 379.

30 BUDNIKOV, P. P. AND BILYAEV, R. A., *Zhur. Priklad. Khim.*, 33 (1960) 1901.
31 WEIR, C. E. AND VAN VALKENBURG, A., *J. Res. Natl. Bur. Standards*, 64A (1960) 103.
32 GELLER, R. F., *et al.*, *ibid.*, 36 (1946) 277; but see also HARRIS, L. A. *et al.*, *Acta Cryst.*, 15 (1962) 615; *J. Amer. Ceram. Soc.*, 45 (1962) 615.
33 MOORE, H. AND MCMILLAN, P. W., *J. Soc. Glass Technol.*, 40 (1956) 66; MENZEL, H. AND ADAM, J., *Glastech. Ber.*, 22 (1949) 237.
34 CHUPKA, W. A., BERKOWITZ, J. AND GIESE, C. F., *J. Chem. Phys.*, 30 (1959) 827.
35 VON WARTENBERG, H., *Z. Anorg. Chem.*, 252 (1943) 136.
36 ZACHARIASEN, A. W., *Z. Phys. Chem.*, 119 (1926) 210.
37 TIEDE, E. AND GOLDSCHMIDT, F., *Ber.*, 62 (1929) 758.
38 LEWIS, J. R., *J. Metals*, 13 (1961) 357; MANNAS, D. A. and SMITH J. P., *ibid.*, 14 (1962) 575; MATYUSHENKO, N. N. *et al.*, *Kristallografiya*, 7 (1962) 862.

CHAPTER 7

Organo-beryllium Compounds

Here we consider those organo-beryllium compounds which contain C–Be bonds. Brief mention is made also of related compounds containing O–Be bonds.

As with the alkali metals, the reactivity of the organo–metallic compounds of the alkaline earths decreases with increasing electronegativity of the metal[1], that is in the order Ba → Be. This gradation is due to the greater degree of covalency of the C–M bond on passing from barium to beryllium. Compounds of the more electropositive elements are ionic in character containing M^{2+} and CR_3^- ions. It is this carbonium anion which is the seat of the high reactivity of the more ionic organo-metallic compounds. It is in accord with these ideas that simple organo derivitives of calcium, strontium and barium have not been obtained; their very high chemical reactivity precludes isolation. The organo derivitives of magnesium and beryllium are much more stable, those of magnesium being the well established Grignard reagents of synthetic organic chemistry. It should be noted that the chemistry of organo-beryllium compounds has been recently reviewed[1a].

1. Dimethylberyllium

This is best prepared on the small scale (1 g) by the interaction of metallic beryllium with dimethylmercury, the mixture being heated to 100° for one to two days[2]:

$$(CH_3)_2Hg + Be = (CH_3)_2Be + Hg$$

The method gives material of 97 % purity, the dimethylberyllium being purified by a vacuum sublimation during which the

References p. 100

mercury is trapped by gold foil[3]. This reaction has also been used[4] for the micro-synthesis of $(CH_3)_2\,^7Be$. For preparation on a larger scale, the Grignard reaction with ethereal beryllium chloride is preferred[5]; the chief difficulty attending the method is to completely remove ether from the dimethylberyllium, a difficulty experienced when preparing other organo-beryllium compounds. As with all organo-beryllium compounds, it is necessary to rigidly exclude air and moisture at all times as dimethylberyllium is spontaneously inflammable in air and violently hydrolysed by water. It decomposes on heating to give $[Be\,CH_2]_n$ as an intermediate and Be_2C as the end product[5a].

Dimethylberyllium forms colourless needles when slowly condensed from the vapour phase. It is not observed to melt; the extrapolated boiling point is 217°/760 mm. The solid has a polymeric structure of infinite chains[6]:

$$\begin{array}{c}CH_3CH_3CH_3\\ \diagdown\!\!\diagup\diagdown\!\!\diagup\diagdown\!\!\diagup\diagdown\!\!\diagup\\ BeBeBeBe\\ \diagup\!\!\diagdown\diagup\!\!\diagdown\diagup\!\!\diagdown\diagup\!\!\diagdown\\ CH_3CH_3CH_3\end{array}$$

It is isomorphous with one of the crystal forms (α') of beryllium chloride which possesses a similar chain structure[7], the compounds contain, respectively, distorted $Be(CH_3)_4$ and $BeCl_4$ tetrahedra. In view of the occurence of a number of polymorphic forms of beryllium chloride (see p. 51), it is possible that different polymorphic forms of dimethylberyllium also exist.

Like beryllium chloride, the vapour of dimethylberyllium contains not only monomers, but also dimers and trimers with higher polymers also present under near saturation conditions[8]. The structures suggested for these are:

$$H_3C-Be-CH_3 \qquad H_3C-Be\!\!\begin{array}{c}\diagup CH_3\diagdown\\ \diagdown CH_3\diagup\end{array}\!\!Be-CH_3$$

$$H_3C-Be\!\!\begin{array}{c}\diagup CH_3\diagdown\\ \diagdown CH_3\diagup\end{array}\!\!Be\!\!\begin{array}{c}\diagup CH_3\diagdown\\ \diagdown CH_3\diagup\end{array}\!\!Be-CH_3$$

The bonding in dimethylberyllium has invoked discussion. The linear monomer presumably involves sp bonds (cf. beryllium chloride monomer, p. 51), but the polymers are usually classed as electron deficient compounds[1,6]. Coates[1] suggests that both carbon and beryllium make use of four tetrahedral (sp^3) atomic orbitals and that three-centre molecular orbitals $Be(sp^3) + C(sp^3) + Be(sp^3)$ are formed from these. Each of these molecular orbitals hold two electrons to give a bent Be–C–Be bond; or, alternatively, each Be–C bond may be regarded as a half bond.

It is relevent to point out the close similarity between dimethylberyllium and beryllium chloride, both as regards structure and chemical reactions. It is possible to formulate crystalline beryllium chloride as a basically ionic compound, but with a Be–Cl bond considerably covalent in character because of the mutual polarization of the ions (*cf*. p. 50-52). Accordingly, it is reasonable to formulate dimethylberyllium similarly and to consider it to contain Be^{2+} and CH_3^- ions, again with a large degree of covalent character due to polarization of the CH_3^- by the Be^{2+} ion. Such a formulation, although crude and elementary, has the advantage of reducing the emphasis on the electron deficient nature of dimethylberyllium (*cf*. beryllium hydride and borohydride, p. 83).

It is of interest that Coates and Green[9] have shown that coloured bipyridyl complexes of the type $BipyBeX_2$ can be prepared:

X in $BipyBeX_2$	colour
Cl	white
Br	pale cream
I	yellow
Ph	yellow
Me	yellow
Et	red

Coloured pyridine derivitives have also been obtained with the undoubtedly electron-deficient decaborane[10]. It is suggested[9] that the transition causing these colours is an electron transfer from one of the Be–X bonds to the lowest unoccupied orbital of bipyridyl. On this view the Be–X bond is acting as an electron donor in the excited state of the complex.

Dimethylberyllium, like the beryllium halides, forms addition complexes. Only molecules with strong donor properties will combine with dimethylberyllium, as the heat of co-ordination must exceed the heat of polymerization of Me$_2$Be. The stability order of these complexes decreases N⟩P⟩O with these as the co-ordination centre of the ligand[11]. This is the order expected when single coordinate links are formed without complications due to double bonding or similar influences of π orbitals. With trimethylamine a volatile compound Me$_2$Be.NMe$_3$ is formed (m. p. 36°, vapour pressure at 36°, 4 mm.), and the compound (Me$_2$Be)$_2$(NMe$_3$)$_3$ is also observed at lower temperatures (−20°)[11]. Dimethylamine reacts with dimethylberyllium to give Me$_2$Be.NHMe$_2$, melting with decomposition at 44° and with the evolution of methane and the formation of (MeBe.NMe$_2$)$_3$[12]. The latter is assigned the cyclic structure:

$$\begin{array}{c} \text{Me}_2 \ \ \text{Me} \\ \text{N}\!-\!\text{Be} \\ \text{Me}\!-\!\text{Be} \ \ \ \ \ \text{NMe}_2 \\ \text{N}\!-\!\text{Be} \\ \text{Me}_2 \ \ \text{Me} \end{array}$$

Weaker donor molecules, such as trimethylphosphine, have a co-ordination affinity for dimethylberyllium only about as great as that of the dimethylberyllium molecules for one another. This leads to the formation of a number of co-ordination complexes with various Be : P ratios. Thus when a bulb containing a little dimethylberyllium and four to five molar proportions of trimethylphosphine is heated from 0–150° a series of phase changes occurs. As the temperature rises the condensed phase alternately melts and freezes some five or six times until only solid dimethylberyllium is left. Coates and Huck[11] suggest the successive formation of a series of compounds of increasing chain length, trimethylphosphine acting as a chain ending group.

$$\begin{array}{cc} \text{H}_3\text{C} \diagdown \ \ \diagup \text{PMe}_3 & \text{H}_3\text{C} \diagdown \ \ \diagup \text{CH}_3 \diagdown \ \ \diagup \text{PMe}_3 \\ \text{Be} & \text{Be} \ \ \ \ \ \ \text{Be} \\ \text{H}_3\text{C} \diagup \ \ \diagdown \text{PMe}_3 & \text{H}_3\text{C} \diagup \ \ \diagdown \text{CH}_3 \diagup \ \ \diagdown \text{PMe}_3 \end{array}$$

Owing to its polymeric nature, dimethylberyllium is insoluble in such liquids as benzene, toluene or phenetole. It dissolves in oxygen containing solvents like ether or tetrahydrofuran, presumably by co-ordination with the solvent. These solutions conduct electricity and beryllium metal can be deposited from them, although not, it appears, in a very pure form[13].

Dimethylberyllium reacts with diborane to give an unstable, easily sublimable solid approximating to $MeBe.BH_4$[2], and further reaction with diborane yields BeB_2H_6 (see p. 83). Dimethylberyllium also undergoes several reactions which are typical of Grignard reagents[5]; for example, it gives $PhMe_2C.OH$ on reaction with benzyl chloride. However, as might be expected from the greater electronegativity of beryllium relative to magnesium, dimethylberyllium is less reactive than Grignard reagents.

2. Diethylberyllium

This is prepared by the reaction of beryllium chloride with ethylmagnesium bromide in ether[14]. As with dimethylberyllium, to completely separate it from ether is difficult, prolonged pumping followed by vacuum distillation gives a product boiling at 63°/0.3 mm. which contains only 2% ether. Solid diethylberyllium is probably associated; this is shown by its low vapour pressure which indicates an extrapolated boiling point of 194°. Dipole moment measurements[15, 16] show that diethylberyllium is monomeric in electron donor solvents such as dioxan, possessing a finite dipole moment, but it is polymeric in solvents such as benzene or heptane, having then a zero dipole moment.

Diethylberyllium reacts with diphenylamine in benzene[17]:

$$(C_2H_5)_2Be + 2Ph_2NH = Be(NPh_2)_2 + 2C_2H_6$$

$Be(NPh_2)_2$ is a white powdery material, insoluble in hydrocarbons and probably polymeric in character. The corresponding aluminium compound, $Al(NPh_2)_3$, has also been prepared.

Diethylberyllium reacts with sodium hydride in boiling ether to give a solution which on evaporation deposits colourless needles of sodium hydrodiethylberyllate, $Na(HEt_2Be)$, m.p. 198°[18]. It is stated to have a dimeric structure which allows the beryllium to achieve a co-ordination number of four:

$$\text{Na}\left[\begin{array}{c}\text{Et}\diagdown\text{Be}\diagup\text{H}\diagdown\text{Be}\diagup\text{Et}\\ \text{Et}\diagup\phantom{\text{Be}}\diagdown\text{H}\diagup\phantom{\text{Be}}\diagdown\text{Et}\end{array}\right]$$

A similar compound is formed from dimethylberyllium and sodium hydride, but it is more soluble in ether. Complex salts of diethylberyllium of the type MX.nBe$(C_2H_5)_2$, where n is 1, 2 and 4, have been recently obtained by the action of alkali fluorides, cyanides and tetraethylammonium chloride on ethereal diethylberyllium[18a]. The addition of beryllium chloride to an ether solution of lithium hydrodiethyl-beryllate causes the pecipitation of lithium chloride. Evaporation of the resulting solution gives a viscous liquid; this is ethyl-beryllium hydride, and is free from lithium or chloride. It is thought to be $Be_3H_2Et_4$, although it could be a mixture of Et_2Be and EtBeH. This route to hydride derivitives of beryllium is being further studied[18].

Diethylberyllium begins to decompose above 65°, rapidly at 190–200°. The course of thermal decomposition is complicated, ethane, ethylene and butane are the principle products, together with smaller quantities of 3-hexene, 1-3-cyclohexadiene and benzene[14]. The residue is a viscous oil (possibly $Be_3H_2Et_4$) readily hydrolysed by water with formation of hydrogen, ethane and ethylene.

3. Di-isopropylberyllium

$(Me_2CH)_2Be$ is prepared from beryllium chloride and *iso*propylmagnesium bromide in ether[19]. This gives a colourless liquid from which the ether can be separated only with difficulty. The ether free material melts at −9.5° and has an extrapolated boiling point of 280°. It is soluble in benzene, in which it is dimeric, and the low vapour pressure of the pure material is also consistent with the liquid being dimeric:

$$\text{Me}_2\text{HC}-\text{Be}\diagup^{\text{CHMe}_2}\diagdown_{\text{CHMe}_2}\text{Be}-\text{CHMe}_2$$

Di-*iso*propylberyllium begins to slowly loose propane at 50°, rapidly at 200°, when a viscous non-volatile polymer of *iso*propylberyllium hydride is formed, $(Me_2CH.BeH)_x$. This is hydrolysed by water to beryllium oxide, propane and hydrogen.

Below are set out the reactions of di-*iso*propylberyllium with amines and methanol[19]:

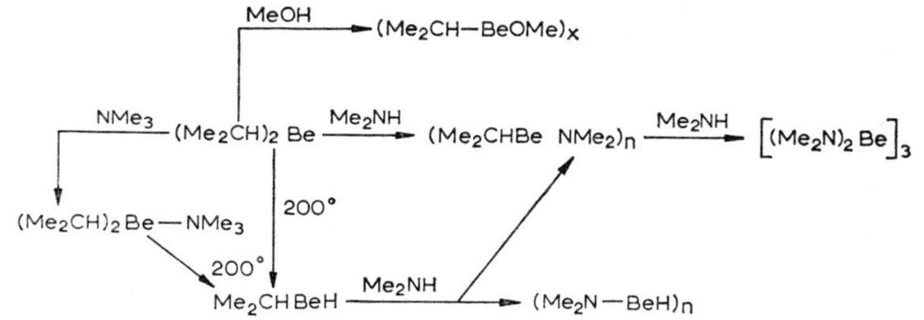

$$\begin{array}{c}
\xrightarrow{\text{MeOH}} (Me_2CH-BeOMe)_x \\
\xrightarrow{NMe_3} (Me_2CH)_2Be \xrightarrow{Me_2NH} (Me_2CHBe\ NMe_2)_n \xrightarrow{Me_2NH} \left[(Me_2N)_2Be\right]_3 \\
(Me_2CH)_2Be-NMe_3 \quad 200° \\
\searrow 200° \\
Me_2CHBeH \xrightarrow{Me_2NH} (Me_2N-BeH)_n
\end{array}$$

The compound $[(Me_2N)_2Be]_3$ is considered to have a cyclic structure:

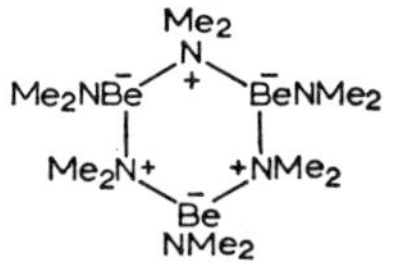

$$\begin{array}{c}
Me_2 \\
N \\
Me_2N\bar{B}e\ +\ \bar{B}eNMe_2 \\
Me_2\overset{+}{N}\overset{+}{N}Me_2 \\
Be \\
NMe_2
\end{array}$$

4. Di-*tert*.-butylberyllium

$(Me_3C)_2Be$ is prepared by the action of beryllium chloride on *tert*.-butylmagnesium chloride in ether[20]. It is less stable than the lower beryllium dialkyls and can be kept satisfactorily only at a low temperature. It has not been prepared free of ether, and decomposition of the material begins above 50°, and is rapid at 200° with formation of isobutene and liberation of ether; the residue is beryllium hydride (see p. 82).

5. Diphenylberyllium

This is prepared by heating beryllium metal with diphenylmercury for 72 h at 150–170° in the presence of a little mercuric chloride as

References p. 100

catalyst[21]. Alternatively, it is obtained by the interaction of beryllium chloride with phenyllithium in ether[13]. It decomposes without melting at 160°, and is more soluble in ether and tetrahydrofuran than in benzene or xylene. The compound is monomeric in oxygenated solvents as shown by its then having a finite dipole moment[16]. This conclusion is supported by conductivity measurements[22] and by the isolation of the compound $Ph_2Be(OEt_2)_2$ (m.p. 28–32°) from ether solution[21]. Diphenylberyllium has a zero dipole moment in heptane[16], which indicates that it is polymeric in this solvent, as it is also in the solid state.

It is interesting that the existence of phenylberyllium bromide and related compounds has been recently denied[23]. No exchange has been found to occur between diphenylberyllium and beryllium bromide containing ^7Be; thus the equilibrium $R_2M + MX_2 \rightleftharpoons$ $\rightleftharpoons 2RMX$ is not established. Some interaction must however take place since, although beryllium bromide forms a two phase system in ether (see p. 57-58), the addition of diphenylberyllium affords a one phase system.

6. *Dicyclopentadienylberyllium*

This compound has been prepared from cyclopentadienylsodium and anhydrous beryllium chloride in ether or benzene[24]. It is monomeric, as shown by its relatively high volatility (it can be purified by vacuum sublimation at 25–45°) and by its possessing a finite dipole moment in both benzene (2.46D) and cyclohexane (2.24D)[25, 26]. There has been some discussion concerning the structure of dicyclopentadienylberyllium, the evidence at present being considered to indicate that the two C_5H_5 rings are not parallel to each other but are inclined at an angle of 90° (ref. 26a).

7. *Catalytic reactions*

It should be briefly mentioned that organo-beryllium compounds act as catalysts for the polymerization of olefins (*e.g.* ref. 27) and of vinyl compounds (*e.g.* ref. 28).

8. Beryllium compounds of the spiran type

Two compounds of this type have been described[29]:

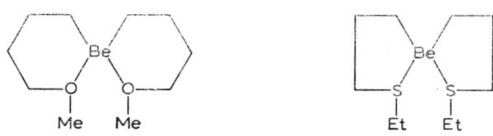

Like all compounds containing C–Be bonds, these are sensitive to air and moisture. They are obtained by addition of $MeO(CH_2)_4Cl$ or $EtS(CH_2)_3Cl$ to a mixture of iodine activated magnesium and beryllium chloride in ether. The compounds are monomeric in benzene.

9. Beryllium alcoholates and phenolates

Beryllium alcoholates are rather difficult to prepare in a pure state. Mixing beryllium chloride and sodium ethoxide in ether gives a precipitate of beryllium ethoxide $Be(OC_2H_5)_2$, in which some of the alkoxy groups are replaced by OH groups[30]. Evidently "hydrolysis" occurs in presence of the alkaline sodium ethoxide. Rather better material is obtained by the reaction of beryllium metal with ethanol[30, 31]:

$$Be + 2C_2H_5OH = Be(OC_2H_5)_2 + H_2$$

This is catalysed by the addition of iodine or mercuric chloride. Beryllium ethoxide is a white amorphous solid which neither melts or sublimes. It is practically insoluble in the usual organic solvents and is decomposed by water. Nothing is known about its structure, which would be of interest for comparison with the structures of the beryllium alkyls and halides.

Beryllium methoxide is less well defined than the ethoxide; Wiberg and Hartwimmer[32] obtained an alcohol-soluble beryllium derivitive, to which they gave the formula $Be(OCH_3)_2$, by heating beryllium borohydride with methanol:

$$Be(BH_4)_2 + 8CH_3OH = Be[B(OCH_3)_4]_2 + 8H_2$$
and $Be[B(OCH_3)_4]_2 = Be(OCH_3)_2 + 2B(OCH_3)_3$

Coates et al.[12] state that dimethylberyllium reacts with methanol to give a dimeric methoxide:

$$MeBe\underset{\underset{Me}{O}}{\overset{\overset{Me}{O}}{\diamond}}BeMe$$

This material shows a strong tendency to disproportionate into dimethylberyllium and beryllium methoxide. A similar compound is obtained by action of methanol on di-*iso*propylberyllium[19].

Beryllium phenolates are somewhat better defined. The phenolate itself is obtained by fusing phenol with beryllium chloride[33], or by action of phenol on the ether complex of beryllium chloride[34,35]:

$$BeCl_2 \cdot 2(C_2H_5)_2O + 2C_6H_5OH = Be(OC_6H_5)_2 + 2HCl + 2(C_2H_5)_2O$$

Compounds with 2-naphthol and *o*-, *m*- and *p*-cresol, catechol, resorcinol, hydroquinone and pyrogallol have been similarly prepared[34-36]. They are all white, hygroscopic materials, which are slowly hydrolysed in air and which on heating give smooth weight-loss curves.

In aqueous solution, salts of the general formula

$$M_2[Be(C_6H_4O_2)_2], xH_2O,$$

where M = NH$_4$, Na, K and Ba, have been obtained by dissolving beryllium hydroxide in alkaline solutions of catechol[37]. These compounds are more closely related to the complex compounds described in Chapter 5 than to the organo-compounds of beryllium here considered.

REFERENCES

1 COATES, G. E., *Organo-metallic compounds*, Methuen, London, 1956.
1a BALUEVA, G. A. AND IOFFE, S. T., *Uspekhi Khim.*, 31 (1962) 940.
2 BURG, A. B. AND SCHLESINGER, H. I., *J. Amer. Chem. Soc.*, 62 (1940) 3425.
3 RABIDEAU, S. W. MOHAMMED ALEI AND HOLLEY, C. E., *U.S.A.E.C. report LA-1687* (1954).

REFERENCES

4 MUXART, R. MELLET, R. AND JAWORSKY, R., *Bull. Soc. Chim. France*, (1956) 445.
5 GILMAN, H. AND SCHULTZ, F., *J. Chem. Soc.*, (1927) 2663; idem, *J. Amer. Chem. Soc.*, 49 (1927) 2904; GILMAN, H. AND BROWN, R. F., *ibid.*, 52 (1930) 4480.
5a GOUBEAU, J. AND WALTER, K., *Z. Anorg. Chem.*, 322 (1963) 58.
6 SNOW, A. I. AND RUNDLE, R. E., *Acta' Cryst.*, 4 (1951) 348.
7 RUNDLE, R. E. J. *Phys. Chem.*, 61 (1957) 45.
8 COATES, G. E., GLOCKLING, F. AND HUCK, N. D., *J. Chem. Soc.*, (1952) 4496.
9 COATES, G. E. AND GREEN, S. I. E., *Proc. Chem. Soc.*, (1961) 376; *J. Chem. Soc.*, (1962) 3340.
10 GRAYBILL, B. M. AND HAWTHORNE, M. F., *J. Amer. Chem. Soc.* 83 (1961) 2673.
11 COATES, G. E. AND HUCK, N. D., *J. Chem. Soc.*, (1952) 4501.
12 COATES, G. E., GLOCKLING, F. AND HUCK, N. D., *ibid.*, (1952) 4512.
13 WOOD, G. B. AND BRENNER, A., *J. Electrochem. Soc.*, 104 (1957) 29.
14 GOUBEAU, J. AND RODEWALD, B., *Z. Anorg. Chem.*, 258 (1949) 162.
15 STROHMEIER, W. AND HUMPFNER, K., *Z. Elektrochem.*, 60 (1956) 1111.
16 STROHMEIER, W. et al., *ibid.*, 63 (1959) 537.
17 LONGI, P., MAZZANTI, G. AND FRANCESCO, B., *Gazz. Chim. Ital.*, 90 (1960) 180.
18 COATES, G. E. AND COX, G. F., *Chem. and Ind.*, (1962) 269.
18a STROHMEIER, W. AND GERNERT, F., *Ber.*, 95 (1962) 1420.
19 COATES, G. E. AND GLOCKLING, F., *J. Chem. Soc.*, (1954) 22.
20 Idem, *ibid.*, (1954) 2526.
21 WITTIG, G., et al., *Ann.*, 571 (1951) 167; 577 (1952) 11.
22 STROHMEIER, W. AND SEIFERT, F., *Z. Elektrochem.*, 63 (1959) 683.
23 DESSY, R .E., *J. Amer. Chem. Soc.*, 82 (1960) 1580.
24 FISCHER, E. O. AND HOFMANN, H. P., *Ber.*, 92 (1959) 482.
25 FISCHER, E. O. AND SCHREINER, S., *ibid.*, 92 (1959) 938.
26 STROHMEIER, W. AND V. HOBE, D., *Z. Elektrochem.*, 64 (1960) 945.
26a FRITZ, H. P. AND SCHNEIDER, R., *Ber.*, 93 (1960) 1171; FRITZ, H. P. AND FISCHER, E. O., *J. Chem. Soc.*, (1961) 547.
27 CRAWFORD, J. W., *British patent* 795,971 (1958).
28 ZUTTY, N. L. AND WELCH, F. J., *J. Polymer Sci.*, 43 (1960) 445.
29 BAHR, G. AND THIELE, K. H., *Ber.*, 90 (1957) 1578.
30 TUROVA, N. YA., NOVOSELOVA, A. V. AND SEMENENKO, K.N., *Zhur. Neorg. Khim.*, 4 (1959) 997.
31 SCHMIDT, J. M., *Ann. Chim. (Paris)*, 11 (1929) 433.
32 WIBERG, E. AND HARTWIMMER, R., *Z. Naturforsch.*, 10B (1955) 290.
33 FRICKE, R. AND HAVERSTADT, L., *Z. Anorg. Chem.*, 140 (1925) 123.
34 SILBER, P., *Ann. Chim. (Paris)*, 7 (1952) 182.
35 TUROVA, N. YA., NOVOSELOVA, A. V. AND SEMENENKO, K. N., *Zhur. Neorg. Khim.*, 4 (1959) 549.
36 PRASED, S. AND SRIVASTAVA, K. P., *J. Ind. Chem. Soc.*, 35 (1958) 263.
37 ROSENHEIM, A. AND LEHMANN, F., *Ann.*, 440 (1924) 153.

CHAPTER 8

The Extractive Metallurgy of Beryllium

1. Beryllium ores

Despite its low atomic number, beryllium is one of the rarer elements; in the earth's crust it has been variously estimated to be present to the extent of 6 p.p.m.[1] and 2 p.p.m.[2]. This low occurrence is ascribed to the ability of beryllium to react with high energy protons and neutrons. In support of this explanation beryllium is a scarce element in solar and stellar atmospheres, but relatively more abundant in cold interstellar matter where such nuclear reactions would not be favoured.

Terrestially, beryllium is an element which was concentrated in the residual magma during the solidification of large molten rock masses, so that it is most commonly found in pegmatite veins and lenticles, the last portions of granite domes to crystallize. This concentration in the residual magmas, often as large crystals of beryl, is of great value as it has rendered the element more accessible than its low concentration in the earth's crust would suggest. Its behaviour is in contrast to that of an element such as gallium which, despite an occurrence in the earth's crust of 15 p.p.m., occurs more rarely in recoverable minerals because it has not been concentrated by any geological process.

In Table 5 are listed some of the principal beryllium minerals. Of these beryl is by far the most abundant, although phenacite, chrysoberyl and bertrandite have been reported as being constituents of recently discovered beryllium bearing deposits in the United States. Barylite and helvine have been observed on occasion, but not yet in sufficient quantities to constitute possible commercial sources of beryllium. At present beryl continues to be the only beryllium ore of commercial significance.

TABLE 5

BERYLLIUM MINERALS

Name	Composition	Crystal system
Beryl	$3BeO \cdot Al_2O_3 \cdot 6SiO_2$	hexagonal
Phenacite	Be_2SiO_4	hexagonal
Chrysoberyl	$BeAl_2O_4$	orthorhombic
Bertrandite	$Be_4Si_2O_7(OH)_2$	orthorhombic
Barylite	$BaBe_2Si_2O_7$	orthorhombic
Helvine	$(MnFeZn)_8Be_6Si_4O_{24}S_2$	isometric

Beryl is mainly recovered from broken pegmatites by hand sorting; by this "cobbing" only relatively large fragments of beryl are recovered. Such methods are applicable to coarse grained deposits, and then only to the weathered surface material. They yield from 20–30% of the beryl in the rock, and considerable efforts have been made to develop methods of mineral dressing for the recovery of beryl of finer grain size. Successful flotation methods have been developed by a number of laboratories and it is now possible to prepare concentrates with 2–10% of beryllium oxide* from a pegmatite rock containing 0.1–0.2% beryllium oxide[3]. Such concentrates consist of mixtures of beryl with other minerals. For example, a flotation concentrate containing 5.4% beryllium oxide consisted of beryl 41, mica 17, spodumene 11, feldspar 8, quartz 8, apatite 5, amblygonite 3 and non-magnetic opaques 5%.

Studies of beryl by X-rays has shown it to contain discrete $Si_6O_{18}^{12-}$ ring ions[4]. These ions are symmetrically surrounded by Be^{2+} ions in four-fold co-ordination and Al^{3+} ions in six-fold co-ordination. The metal ions are completely shielded by the silicate rings, which explains the chemical inertness of the mineral and also the difficulties experienced in floating it. As the $Si_6O_{18}^{12-}$ ring-ions are packed one above another in the crystal, tunnels exist down the centre of these rings which enable the crystal to accomodate small molecules such

---

* It is normal commercial practice to state the beryllium content of ores in terms of beryllium oxide. A similar terminology is often employed in analytical chemistry.

as helium or water. Natural beryl usually departs markedly from the ideal formula given in the above table. It generally contains 10–12% beryllium oxide which is appreciably less than the theoretical value of 13.96%. Alkali metals are very common constituents of natural beryl although the place they occupy in the crystal structure is debatable[5].

2. *Extraction of beryllium from beryl*

There are now two recovery processes in commercial operation. In the first, beryllium is selectively extracted by roasting beryl with sodium fluorosilicate and leaching with water (Copaux process). In the second the beryl is rendered soluble by heat treatment, the modified material is decomposed by sulphuric acid, and the beryllium and other soluble sulphates are dissolved in water. In contrast to the Copaux process this break-down process is quite unselective, and requires the separation of beryllium and aluminium to be undertaken in the subsequent stages.

3. *Copaux fluorosilicate process*

This process stems from that devised by Copaux[6] and consists in heating beryl with sodium fluorosilicate at 700–750° at a weight ratio of sodium fluorosilicate to beryl of 0.8–1 : 1[7, 8]. The roast material is extracted with water when beryllium, as fluoride or sodium fluoroberyllates, is dissolved, whilst the aluminium remains as insoluble chiolite, Na_2AlF_5, or cryolite, Na_3AlF_6. This procedure is highly selective for beryllium and the leach liquors obtained are remarkably free of anything except sodium, beryllium and fluorine. It is possible to somewhat increase the amount of beryllium dissolved at this stage, with only a slight increase in the dissolved impurities, by leaching with water adjusted to pH 4 with sulphuric acid[9].

After the beryllium fluorides have been leached from the roast product, the beryllium is precipitated by bringing the liquor to pH 12 with alkali:

$$Na_2BeF_4 + 2NaOH = Be(OH)_2 + 4NaF$$

The solubility of beryllium in the presence of an excess of fluoride is at a minimum at this acidity. Thus precipitated in the cold, beryllium hydroxide is gelatinous, but it separates as a granular, easily filterable material from hot solutions[10]. A purer product is obtained using the beryllate process described by Bryant[11]. In this crude beryllium hydroxide precipitated at pH 12 is re-dissolved in the minimum quantity of sodium hydroxide to form sodium beryllate. The solution is slowly heated to 45–50° to coagulate the impurities, principally ferric hydroxide, insoluble in alkali. These are removed by filtration. Granular beryllium hydroxide is precipitated by boiling the clarified beryllate solution.

The most important part of the Copaux process is the roasting step. Hyde, Robinson, Waterman and Waters[12] suggest that the reaction occurs in three stages. In the first the beryl is considered to make available the ions Be^{2+}, Al^{3+}, Si^{4+} and O^{2-}, and sodium fluorosilicate the ions Na^+, F^- and SiF_6^{2-}. The proportion of sodium fluoride to fluorosilicate is low and little silicon tetrafluoride has yet been lost. Early product species are sodium fluoroberyllate, $NaBeF_3$, cryolite, Na_3AlF_6 and α–cristobalite, SiO_2. A notable feature of this stage is the high proportion of the total beryllium which is rendered water soluble in a comparatively short period e.g. 62% and 70% in 15 min. at 750 and 800° respectively. It should be noted that sodium fluorosilicate is itself decomposed at the roasting temperature of 750°:

$$Na_2SiF_6 \rightleftharpoons 2NaF + SiF_4$$

The pressure of silicon tetrafluoride above sodium fluorosilicate is 65 mm at 530° and 560 mm at 672°[14].

During the second stage of the reaction the proportion of sodium fluoride to fluorosilicate is higher and this change is accompanied by the appearance of albite, $NaAlSi_3O_8$. The third stage is marked by a progressive decrease in the proportion of water extractable beryllium, although no beryllium is lost from the roast. A decrease also occurs in the amount of cryolite and α-cristobalite and an increase in the albite and sodium fluoride. It is thought that the non-extracable beryllium has been incorporated into the albite structure. This third stage of the reaction is entirely disadvantageous.

Robinson et al.[12] suggest the following tentative mechanism for the Copaux roasting reaction. Initially a molten or semi-molten film of a sodium fluoride–beryl flux is formed on the mineral particles. Through this film Na^+ and F^- ions pass inwards and O^{2-} outwards, the first two to form $NaBeF_3$ and Na_3AlF_6 and the O^{2-} to react with the SiF_6^{2-} ions to form SiO_2 and F^-. Further reaction proceeds by the interfacial film moving inwards on the surface of the residual beryl. The three reaction stages are shown schematically:

First Stage

$$Na^+ + Be^{2+} + 3F^- = NaBeF_3 \text{ (glass)}$$

$$3Na^+ + Al^{3+} + 6F^- = Na_3AlF_6 \text{ (crystalline)}$$

$$Si^{4+} + 2O^{2-} = SiO_2 \text{ (crystalline)}$$

Second Stage

$$Na^+ + Al^{3+} + 3Si^{4+} + 8O^{2-} = NaAlSi_3O_8 \text{ (crystalline)}$$

Third Stage (in absence of beryl)

$$Na_3AlF_6 + 4SiO_2 = NaAlSi_3O_8 \text{ (cryst.)} + 2NaF \text{ (cryst.)} + SiF_4$$

As the beryllium fluorides formed on roasting are glassy and X-ray amorphous, it is impossible to be at all certain which fluoride is formed in the reaction; a mixture of BeF_2, NaF, $NaBeF_3$ and Na_2BeF_4 may be present.

The work of Robinson et al.[12] indicates the need to conserve silicon tetrafluoride in the roasting process, as its removal from the system increases the amount of sodium fluoride produced. Although sodium fluoride attacks beryl, it renders only about 25% of the beryllium water soluble and also causes the undesirable formation of albite. The work indicates the need to keep the roasting time to a minimum. All the available beryllium is liberated quite rapidly, and there is an actual loss of water soluble beryllium on prolonged roasting. The latter observation is of importance as it has been the practice in some industrial operations to employ roasting times of up to 8 hours. This procedure has been dictated by the design of roasting furnace and the use of relatively large briquettes of beryl and fluorosilicate. The low thermal conductivity of the briquettes requires that they have a long residence time in the furnace before they are heated throughout to 750°; this leads to the outside of the briquettes being overheated and to a fall in the soluble beryllium.

The operation of the Copaux process with lower grade flotation concentrates is seriously hindered by the presence of both phosphate and calcium containing minerals[9, 13]. Phosphate interferes by reducing the amount of water soluble beryllium, produced on roasting, although the unextracted beryllium can be recovered by leaching with dilute sulphuric acid (pH<2.5) when beryllium phosphates pass into solution. Unfortunately other impurities, in particular calcium and silica, are also dissolved. These seriously interfere with the operation of the beryllate process by combining with beryllium hydroxide to form alkali insoluble materials.

The reason for the interference of calcium in the Copaux process is less clear, but its presence certainly reduces the amount of water soluble beryllium obtained on roasting. Possibly sparingly soluble calcium fluoroberyllate is produced. For satisfactory operation of the Copaux process with lower grade concentrates it is necessary to remove the calcium and phosphate containing minerals at the ore-dressing stage. If this is not practicable, the concentrate must be pre-leached with cold 2 N-hydrochloric acid before roasting with sodium fluorosilicate, since apatite and amblygonite, the most common phosphate and calcium bearing minerals in the concentrates, are soluble in dilute hydrochloric acid[13]. Sometimes it is

References p. 116

necessary to heat the concentrate to about 750° before pre-leaching to obtain a satisfactory removal of calcium and phosphate. Calcium fluoride (fluorite) is a more serious impurity as it cannot be readily removed from the concentrate by such acid treatment[9].

Rather similar results to those of Robinson et al.[12] as to the course of the Copaux roasting reaction have been found in experiments with lower grade concentrates[13]. When the reaction is undertaken in a sealed furnace at 750°, using a concentrate containing 2% beryllium oxide at a reagent: concentrate weight ratio of 0.5 : 1, a pressure maximum occurs after approximately four minutes, at which time about 60% of the sodium fluorosilicate is decomposed. This pressure increase is much sharper than when sodium fluorosilicate is heated alone. Production of water soluble beryllium is virtually complete after one hour, re-adsorption of the silicon tetrafluoride being then nearly complete. The presence of silicon tetrafluoride increases the initial reaction rate possibly because it maintains for a longer period a high concentration of fluorosilicate. A significant feature of this work is that it shows silicon tetrafluoride to play a direct part in the roasting process. The attainment of maximum beryllium water solubility depends on the re-adsorption of silicon tetrafluoride. These results again emphasize the need to retain silicon tetrafluoride in the system and that relatively short roasting times can be employed.

It is important to note that the concentration of beryllium which can be achieved in the water leach liquor, is dependent upon the amount of fluoride dissolved *i.e.* upon the F : Be mole ratio in solution. A study of the NaF–BeF_2–H_2O system[15] has shown that when this ratio is 4 : 1, the solubility of beryllium amounts to 2.7 g BeO per litre. The solubility is reduced as the F : Be ratio in solution rises above 4 and is increased when the ratio falls below 4. Thus a leach liquor containing 5 g BeO per litre had an F : Be mole ratio of 3.2 : 1. Such solubility considerations place a further premium upon the conservation of silicon tetrafluoride during roasting. Not only does conservation reduce the amount of sodium fluoride formed by decomposition of sodium fluorosilicate, but less sodium fluorosilicate is required to obtain a given beryllium extraction.

4. Sulphate routes for recovery of beryllium from beryl

The best known procedure of this type is the fuse–quench process operated by the Brush Beryllium Co. with high grade beryl ($>$10% beryllium oxide) as feed. The beryl is melted in an electric furnace at about 1600–1650° and quenched by pouring through a high velocity water jet into a tank of water. The resulting glass is reheated to 900–950° (post-heating) for a short time to further increase its chemical reactivity. It is finely ground, treated with a controlled excess of sulphuric acid (sulphation), and heated to 250–300° (curing) in order to insolubilize the silica. The beryllium and other soluble sulphates are leached out with water.

The strength of acid employed for sulphation of the products of the fuse–quench process is important[9,13]. Well crystallised material, such as that obtained on post-heating, requires sulphuric acid of a concentration 85 wt.% for optimum beryllium extraction. For glasses amorphous to X-rays obtained by the rapid quenching of molten beryl, the best beryllium extractions are obtained with more dilute acid (20–40%). Even so, the proportion of beryllium extracted from such glasses is usually lower than that obtained from the partly crystallised post-heated materials.

The increased reactivity of the post-heated glass is due to the crystallization of chrysoberyl ($BeAl_2O_4$), beryllia, mullite ($3Al_2O_3 \cdot 2SiO_2$) and a new phase which has been provisionally called modified phenacite[9,13,16]. The nature of this latter phase is still uncertain, but the latest X-ray data indicate that it is probably a mixed beryllium-aluminium orthosilicate related both to mullite and to phenacite.

To obtain information on the mechanism of the fuse–quench process, studies have been made of the phase changes which are observed during the cooling of pure beryl melts[16]. High-temperature microscopy has been used, supported by X-ray investigation of the cooled samples. When heated, beryl begins to sinter at about 1300°, the first signs of liquid appear at 1490° and clear, stable melts are formed at 1590°. The phases which separate on cooling depend very much upon the conditions. With a relatively slow cooling rate

References p. 116

(<200° per sec) the primary phase is beryllia and, in melts kept at about 1500°, its crystallization is followed by that of chrysoberyl. These two phases are also the first to appear from slowly heated beryl glasses. Both these phases are based upon the close-packed stacking of oxygen ions. Faster cooling rates (200–350° per sec) results in the crystallisation of mullite with its more open structure, together with beryllia. With still faster cooling rates (400–500° per sec) only modified phenacite is observed, and quenching rates above 750° per sec produce amorphous glasses. In line with this, the more rapid heating of amorphous glasses is found to cause the crystallization of mullite and modified phenacite in addition to beryllia and chrysoberyl.

It had not been considered possible to apply the fuse–quench process to concentrates containing less than 10% beryllium oxide. However, it has been recently observed[9, 16a] that the addition of small quantities of lime to a low grade concentrate allows the beryllium in the quenched glass to be leached by dilute sulphuric acid. The weight ratio SiO_2: total alkali ($Na_2O + CaO$) should be about 2.5 : 1. By this means it is possible to treat 2–10% beryllium oxide concentrates by the fuse–quench process without a post-heating stage. Sometimes the alkali already in the concentrate fulfils these conditions and obviates the need for lime addition.

Beryl is completely broken down chemically by heating with the alkalies lime (often as limestone) or sodium carbonate (refs.7, 17 and 18). Such processes differ from that just described in using larger quantities of alkali. Although the high alkali processes are not employed in the treatment of high grade beryl they are likely to prove useful for lower grade concentrates[13]. It has been suggested that the heating with limestone be at sintering temperature (1100–1250°)[17], but this is not easy because fusion may occur. When this happens, cooling gives a glass which is difficult both to remove from the container and to grind. It is much better to completely fuse the charge and quench it by pouring into water; this gives a friable, reactive product. Fusion temperatures depend upon the weight ratio of limestone to concentrate. The lowest effective ratio of 0.54 : 1 requires a temperature of about 1500°, but a 1 : 1

mixture melts by 1250°. Since it is the CaO : SiO$_2$ ratio which is significant, and since the proportion of SiO$_2$ in both concentrate and high grade beryl is similar, these limestone: concentrate ratios apply, irrespective of the beryllium content of the concentrate. As in the fuse–quench process, the lime–fused material is sulphated and cured, and the soluble metal sulphates are leached out with water.

The solutions obtained by water leaching the sulphated and cured materials contain up to 10–12 g BeO per litre, together with all the aluminium originally in the concentrate. The bulk of the aluminium in these is separated from the beryllium by crystallization as ammonium alum, (NH$_4$)$_2$SO$_4$.Al$_2$(SO$_4$)$_3$.24H$_2$O, following addition to the solution of ammonium sulphate (or ammonia which forms ammonium sulphate with the excess of sulphuric acid in the liquor). An excess of caustic soda is then added to the filtrate to convert the remaining aluminium and beryllium to aluminate and beryllate, respectively, a little chelating compound, such as E.D.T.A., is present to keep iron iron-like ions in solution. The resulting solution is boiled to precipitate granular beryllium hydroxide. This procedure is at present applied to the solutions derived from the treatment of high grade beryl and works well. The difficulties of operating it with liquors derived from lower grade concentrates[13], are mainly due to the presence of impurities such as phosphate.

Solvent extraction with cation–exchange type solvents appeared to offer an attractive procedure for the recovery of beryllium from sulphate liquors[13, 19, 20]. Many solvents have been tested as possible beryllium extractants and, taking into account the commercial availability of the ester, di-(2-ethylhexyl) phosphoric acid (EHP) appears to be the most suitable for the selective recovery of beryllium from sulphate solutions[19]. It is employed as an 0.5 M solution in kerosene, with the addition of about 4% capryl alcohol as phase modifier. The currently preferred pH of the aqueous phase is 0.5–1.0[19], the fall in the rate of extraction as the pH is reduced to this value (cf. p. 17-19) is more than offset by the greater selectivity of the extractant under these conditions; the removal of bivalent beryllium is increased and that of trivalent aluminium reduced (cf. refs. 19 and 21).

The solvent extraction circuit for beryllium will differ markedly from the well established EHP–uranium circuit. The relatively slow rate of beryllium extraction will require a greater number of stages than the uranium circuit for a $>98\%$ extraction. Moreover, the high molar concentration (about 0.3–0.4 M) of beryllium in the aqueous phase presents difficulties. It is not possible to have an organic continuous phase in the mixers because phase disengagement is then difficult. This makes it difficult to employ an organic: aqueous flow ratio >1 which is needed to counteract the high aqueous beryllium concentration. Nor is it possible to employ EHP solutions >0.5 M, owing to the high viscosity of the organic solvent phase. Thus it is apparent that the flow arrangements employed for beryllium extraction will not be as straightforward as the counter-current methods employed in uranium systems. For a fuller discussion of these aspects of the solvent extraction of beryllium see ref. 19.

A solution containing BeO 3.11, Al_2O_3 23.1 g per litre, total sulphate 1.0 M and an initial pH of 1.0, was extracted with 0.5 M EHP in kerosene with 4% capryl alcohol as phase modifier[13]. A 1 : 1 organic : aqueous flow ratio was employed with seven extraction stages and a residence time of 20 minutes in each mixer. A recovery of 97% of the beryllium was achieved and the weight ratio Al_2O_3 : BeO in the organic phase was 0.225 compared with approximately 7 : 1 in the aqueous feed. Other tests indicate that EHP extraction affords good separations of beryllium from phosphate, lithium and bivalent ions such as Ca^{2+}, Ni^{2+}, or Mn^{2+}. Ferric iron is however extracted preferentially to beryllium, as would be expected from the high affinity of EHP for ferric iron.

Beryllium can be stripped from the organic phase by a solution of caustic soda, ammonium fluoride, or bifluoride; the equilibrium is displaced in favour of the aqueous phase by the formation of beryllate or fluoroberyllate anions. Ferric hydroxide or ferric fluoride and aluminium fluoride are precipitated during stripping, but the formation of moderate amounts of these solids does not cause serious difficulties provided the settler units are of the correct design. Beryllium is recovered from the beryllate solution as hydroxide by boiling, and from the fluoride solution by the crystallization of ammonium fluoroberyllate followed by its thermal decomposition.

5. Chlorination methods

It would appear from the wide differences in their boiling points ($BeCl_2$ 488°, $AlCl_3$ 183°, $FeCl_3$ 319° and $SiCl_4$ 57°) that the fractional distillation of the chlorides would provide a relatively simple method for the separation of beryllium from the other elements occuring in beryl. It has been shown by several workers that the initial chlorination of the beryl is practically feasible, both for high grade beryl[22, 23] and for lower grade concentrates[24]. It can be performed by passing chlorine over an intimate mixture of carbon and powdered beryl at 800–1000°, with the formation of beryllium, aluminium and silicon chlorides. Alternatively, the beryl (or a lower grade concentrate) is heated with carbon to 1400–1750°, when about 50% of the silicon is volatilised as silicon monoxide, and the beryllium, aluminium and the rest of the silicon are converted to carbides. The chlorination of this material at 900° volatilises ⟩ 95% of the beryllium as $BeCl_2$[24].

Despite the wide differences in boiling point, it has not yet been possible to effect a clean separation of beryllium and aluminium by fractional condensation of their chlorides. Even resublimation of the mixture does not always give a satisfactory separation. The reasons for these difficulties are not yet known, but there is sufficient evidence to suggest that they can be overcome. However, when considering the economic feasibility of the process it is necessary to take into account the high cost of chlorine and the doubtful market for the by-product chlorides.

6. Preparation of metallic beryllium

Although many processes for the production of metallic beryllium have been described in the patent and technical literature, only two of these are of present industrial importance. They are the magnesium reduction of beryllium fluoride and the electrolysis of beryllium chloride in a molten chloride electrolyte.

7. Magnesium reduction of beryllium fluoride

The beryllium fluoride for this process must be of high purity as there is little chance of removing impurities during reduction. This compound is prepared by the thermal decomposition of ammonium fluoroberyllate, $(NH_4)_2BeF_4$ (see p. 39). The latter is made by dissolving beryllium hydroxide in ammonium bifluoride solution:

$$Be(OH)_2 + 2NH_4HF_2 = (NH_4)_2BeF_4 + H_2O$$

Beryllium metal and beryllia scrap are also used as feed materials for the above reaction. The ammonium fluoroberyllate solution is purified by the successive addition of (a) calcium carbonate, (b) lead dioxide and (c) ammonium polysulphide. These reagents precipitate respectively (a) aluminium (as hydroxide), (b) manganese and chromium (as manganese dioxide and lead chromate), and (c) copper, lead and nickel (as sulphides). After filtering off these impurities the solution is evaporated to give crystals of ammonium fluoroberyllate.

The beryllium fluoride is reduced when heated with magnesium:

$$BeF_2 + Mg = Be + MgF_2$$

The reaction proceeds rapidly at temperatures as low as 900°, but it is necessary to go above the melting point of magnesium fluoride and metallic beryllium (about 1300°) in order to allow the reaction products to separate. The reduction is strongly exothermic, so that the process requires careful control. Less than the stoichiometric quantity of magnesium is employed (usually about 75%), as the presence of an excess of beryllium fluoride allows the slag to rapidly break up in water and release the particles of metallic beryllium.

The carbon crucible in which reduction is carried out is kept at temperature for 3 hours, after which the cooled charge is wet-milled to separate the beryllium pebbles from the magnesium fluoride slag. The slurry leaving the mill goes to a settling tank and the supernatent liquor, which contains beryllium fluoride, is returned to the mill and eventually to the purification circuit. A little finely divided metallic beryllium passes out of the mill and settles with the magnesium fluoride. It is recovered by leaching the magnesium fluoride with hydrofluoric acid, and recycling the resulting solution to the purification stage as make-up for the ammonium bifluoride solution[25].

The beryllium thus produced contains up to 1.5% magnesium, and must be purified by vacuum melting. Under correct operating conditions the finished metal contains less than 500 p.p.m. of magnesium.

8. Electrolysis of beryllium chloride

The beryllium chloride used in this process is prepared by the chlorination of beryllia (see p. 50):

$$BeO + C + Cl_2 = BeCl_2 + CO$$

The beryllia is itself obtained by the thermal decomposition of beryllium hydroxide prepared by either the Copaux process or by the sulphate route. This chlorination reaction provides a certain degree of purification of the beryllium (*e.g.* silicon is removed as $SiCl_4$), whilst a further purification can take place during the electrolysis.

The electrolyte usually consists of a mixture of equal parts of sodium and beryllium chlorides, although a solution of beryllium chloride in a lithium chloride–potassium chloride eutectic has been recently employed[26, 27]. The cathode is usually nickel or iron, and graphite or iron have been used as anode materials. The operating temperatures are about 350–400° for the $NaCl–BeCl_2$ electrolyte and 450–550° for the $LiCl–KCl–BeCl_2$ electrolyte. The beryllium flake produced by electrolytic methods is of very good quality and, in particular, it can have a low oxygen content. The method is in industrial operation, although not so widely used as is the thermal reduction of beryllium fluoride.

References p. 116

REFERENCES

1. GOLDSCHMIDT, V. M., *Geochemistry*, Oxford University Press, 1954.
2. SANDELL, E. B., *Geochim. Cosmochim. Acta*, 2 (1952) 211.
3. MOIR, D. N. et. al., *Paper 44J, 6th International Mineral Processing Conference, Cannes*, 1963.
4. BRAGG, W. L. AND WEST, J., *Proc. Roy. Soc.*, 111 (1926) 691.
5. FRANK-KAMENETSKII, V. A., *Geokhimiya*, 4 (1959) 912; BELOV, N. V., *ibid.*, 4 (1959) 813.
6. COPAUX, M. H., *Compt. Rend.*, 168 (1919) 610.
7. LUNDIN, H., *Trans. Amer. Inst. Chem. Engrs.*, 61 (1945) 671.
8. KAWECKI, H. C., in Chapter IV Part B. *The Metal Beryllium*, Ed. by WHITE D. W. AND BURKE, J. E., The American Society for Metals, Cleveland 1955.
9. *Unpublished observations*, The National Chemical Laboratory.
10. HIGBIE, K. R. AND FARMER, M. C., *Chem. Eng. Prog.*, 54 (1958) 51.
11. BRYANT, P. S., in *Extraction and Refining of the Rarer Metals*, p. 310. Inst. Min. Metall., London, 1957.
12. HYDE, K. A., ROBINSON, P. L., WATERMAN, M. J. AND WATERS, J. M., *Bull. Inst. Min. Metall.*, 70 (1960-61) 397; *J. Inorg. Nucl. Chem.*, 19 (1961) 237.
13. EVEREST, D. A. AND NAPIER, E., *Paper 17D, 6th International Mineral Processing Conference, Cannes*, 1963.
14. CAILLAT, R., *Ann. Chim. (Paris)*, 20 (1945) 368.
15. VOROB'EVA, O. I. et al., *Zhur. Neorg. Khim.*, 1 (1956) 516.
16. MERCER, R. A. AND MILLER, R. P., *Scientific Report NCL/AE 213*, The National Chemical Laboratory, 1962; *Nature*, 197 (1963) 683.
16a. NAPIER, E. AND DERRY, R., *British Patent Application,*, No 42,039/62.
17. RUNKE, S. M. AND RILEY, J. M., *U.S. Bureau of Mines Report of Investigations 5339*, 1957.
18. CATTRALL, R. W., *South Australian Dept. of Mines, Report. R. D. 86*, 1958.
19. WELLS, R. A., EVEREST, D. A. AND NORTH, A. A., *Nuclear Science and Engineering*, (1963) in the press.
20. DANNELBERG, R. O., BRIDGES, D. W. AND ROSENBAUM, J. B., *U.S. Bureau of Mines Report of Investigation 5941*, 1962.
21. IRVING, H. M., Quart. Rev., 5 (1951) 200.
22. WINTERS, R. W. AND YNTEMA, L. F., *Trans. Amer. Electrochem. Soc.*, 55 (1929) 205.
23. MCTAGGART, F. K., *J. Council Sci. Ind. Res. (Australia)*, 20 (1947) 564.
24. MAY, J. T. AND HOATSON, C. L., *U.S. Bureau of Mines Report of Investigations 6037*, 1962.
25. SCHWENZFEIER, C. W., in Chapter IV, Part C of ref. 8.
26. WONG, M. M., CAMPBELL, R. E. AND BAKER, D. H., *U.S. Bureau of Mines Report of Investigations 5959*, 1962.
27. WINDECKER, C. E., in Chapter IV Part D. of ref. 8.

CHAPTER 9

The Analytical Chemistry of Beryllium

1. Introduction

It is not intended to give an exhaustive review of the analytical chemistry of beryllium such as is provided in ref. 1, or to deal with experimental procedures in detail. Instead attention will be focused on the essentials of those methods which are considered to be of greatest utility.

Beryllium belongs to the ammonium hydroxide analytical group; it is completely precipitated by ammonia at pH 8.5 along with the other members: iron, aluminium, chromium. As usual, fluoride, silica, organic matter and phosphate should all be absent. Beryllium is best separated from aluminium, with which it is commonly associated, by means of 8-hydroxyquinoline; aluminium is precipitated from an acetate–buffered acetic acid solution of pH 5.7, whilst beryllium is only precipitated by this reagent when it is alkaline. Beryllium does not form a precipitate with cupferron, so that it can be separated from metals such as iron or titanium which are precipitated by this reagent. Iron, chromium, nickel, cobalt, tin, molybdenum, zinc and lead can all be separated from beryllium by electrolysis at a mercury cathode, when the beryllium is left in solution.

2. Breakdown of beryllium-containing samples

The complete dissolution of solids containing beryllium is often difficult, especially so with minerals. It is usually effected by fusion with sodium carbonate, potassium hydroxide or sodium fluoride,

References p. 131

friting with sodium peroxide or low temperature fusion with ammonium fluoride. When finely powdered, beryllium minerals can be dissolved by prolonged treatment by hydrofluoric acid with nitric or perchloric acids.

Fluoride fusions have the advantage of quickly and completely removing silica as silicon tetrafluoride, but their use usually involves the removal of fluoride by fuming with sulphuric acid. General laboratory experience[2] shows that fusion with sodium fluoride, or friting with sodium peroxide are the best methods for dissolving beryllium minerals. Potassium hydroxide fusions are best for beryllium oxide and similar products based on it which have been heated to a high temperature ($>1500°$). Ammonium fluoride fusions, which always require a large excess of flux (25 g NH_4F per 5 g sample), are of use in geochemical prospecting, where field operation imposes the use of low fusion temperatures. Health check samples are most simply dissolved by fusion with potassium bisulphate. This procedure only incompletely attacks beryl, but, because beryllium in this form is considered non-toxic, the method does not thereby suffer.

3. Quantitative procedures for beryllium determination

(a) Gravimetric methods

These are employed when accuracy takes precedence over speed and ease of analysis, for instance in the accurate determination of beryllium in a high grade ore.

Two main procedures are employed, based either on the precipitation of beryllium hydroxide[3-5] or beryllium ammonium phosphate[6]. The precipitate is ignited and then weighed as BeO or $Be_2P_2O_7$, respectively. The phosphate has the advantage of giving 3.838 times as much mass per equiv. of beryllium as the oxide. Both procedures make use of the fact that although ethylenediaminetetra-acetic acid (E.D.T.A.) forms strong complexes with many metals, that with beryllium is of low stability (see p. 69). It is possible, therefore, to selectively precipitate beryllium hydroxide in presence of this reagent. As in all precipitations of beryllium from aqueous solutions fluoride must be absent as it strongly complexes beryllium and prevents complete precipitation.

For solutions relatively free from interfering ions, precipitation of beryllium hydroxide at pH 8.5 with ammonia in absence of complexing agents, followed by ignition to 1100–1200° in a platinum crucible, is satisfactory. Ammonia is preferred to sodium hydroxide as co-precipitated sodium ions are not driven off on ignition but remain and contaminate the beryllium oxide finally produced. For solutions of greater impurity, such as are obtained in the analysis of minerals, more precautions are necessary. Describing the analysis of beryl, Brewer[4] says that, after the precipitation of beryllium hydroxide in presence of E.D.T.A. (added as the ammonium salt), the ignited precipitate should be fused with sodium carbonate and the cooled melt leached with water to remove aluminium or phosphate. The residue is ignited and weighed as BeO.

In general, results from the hydroxide method tend to be high owing to absorption of impurities by the beryllium hydroxide. Indeed, the procedure recommended by Brewer succeeds only because the errors compensate one another. There is a small loss of beryllium owing to solubility at the precipitation stage and a rather larger one during the leaching of the sodium carbonate fusion. However, these losses are counterbalanced by impurities (sodium, silica, aluminium) in the oxide after its last ignition[7].

In the phosphate method of Hure et al.[6], $BeNH_4PO_4$ is precipitated at pH 5.2 in presence of the ammonium salt of E.D.T.A., the phosphate being added as ammonium dihydrogen phosphate to the solution buffered by means of ammonium acetate. For the highest accuracy the precipitate is redissolved in dilute hydrochloric acid, reprecipitated, ignited at 1000° and weighed as $Be_2P_2O_7$. The alkali metal ions in the solution should be kept to a minimum as these contaminate the precipitate. They give a "fused" appearance to the ignited $Be_2P_2O_7$, and cause it to stick to the sides of the platinum dish[2]. A very significant advantage of the method is that it does not require phosphate to be separated from the solution before the precipitation of beryllium.

References p. 131

An interesting new gravimetric procedure is that described by Pirtea and Constantinescu[8]. In it, beryllium is selectively precipitated by hexamminocobaltic chloride as the complex $[Co(NH_3)_6]$ $[(H_2O)_2Be_2(CO_3)_2(OH)_3].3H_2O$, and is weighed as such. Satisfactory results are obtained with pure beryllium solutions, but great care is needed during the drying of the precipitate to achieve constant composition.

A brief examination of the application of this method to ore analysis has given low results[7]. Possibly phosphate present in the ore was precipitated as beryllium phosphate; this would have a lower precipitate : beryllium ratio than the cobaltammine complex. An indirect titration of the beryllium in the precipitated complex with E.D.T.A.[9] might overcome errors caused by variation in the composition of the precipitate. However, this apparently elegant method has shown some uncertainty in operation, which so far has prevented its general use[2].

(b) Volumetric methods

Only one reasonably satisfactory volumetric procedure for beryllium exists, and that is both empirical in character and applicable only to relatively pure solutions. Beryllium is first precipitated with ammonia (pH 8.5) from a sulphate or chloride solution and an excess of alkali fluoride is added as potassium or sodium fluoride. Almost quantitative reaction occurs:

$$Be(OH)_2 + 4F^- = BeF_4^{2-} + 2OH^-$$

The amount of alkali liberated is proportional to the beryllium present and is titrated with standard acid[3, 10, 11]. Aluminium reacts in a similar manner and must either be absent, or a correction made for its presence. The same is true of zirconium, rare earths, uranium and thorium which should also be absent. As the reaction is not stoichiometric, all the operations must be precisely standardized and the acid used must be standardized by an exactly similar method. In particular, the titration must be performed slowly, so as to give sufficient time for the reaction to reach equilibrium after each addition of reagent.

The method is useful in practice because of its simplicity, speed and reasonable accuracy and should be of value for the determination of beryllium in relatively pure solutions. The low equivalent weight of beryllium, a handicap in gravimetric work, is an advantage in volumetric analysis as it causes large titres to be given by relatively small amounts of beryllium.

(c) Colorimetric and fluometric methods

Very large numbers of colorimetric or fluometric reagents have been described for the detection and determination of beryllium (for a list see ref. 1). Of these reagents zenia[12] (*p*–nitrophenylazoorcinol) and, in particular, berillon[13] [the tetra-sodium salt of 8-hydroxynaphthalene-3,6-disulphonic acid-(1-azo-2′)-1,8′-dihydroxynaphthalene-3′,6′-disulphonic acid] have proved the most important. Although zenia is widely used for beryllium analysis its sensitivity is rather low, the colour intensity of the complex is temperature dependent (indicating approx. 2.5% increase per degree) and the conditions in which the reagent blank is found are very critical and must be carefully controlled.

In practice berillon seems to be the most satisfactory absorptiometric reagent for beryllium[7]. In strong alkali berillon is magenta to violet depending upon its concentration, and its beryllium complex is blue. Karanovich[13] increased the selectivity of the reagent by adding E.D.T.A. to complex most other metals and ascorbic acid to reduce iron to the non-interferring, ferrous state. However, it is doubtful whether these reagents can effectively eliminate interference due to ferric iron in the strongly alkaline solutions employed. Other possible sequestering agents for iron have been examined, and it is recommended that Nervanaid F [N,N-bis-(2-hydroxyethyl) glycine] be used in place of ascorbic acid when much iron is present[7]. The solution at the time of measurement should contain not more than 16 µg Be per 50 ml. Its absorbency is measured at 630 mµ (Unicam SP 500 spectrophotometer-red photocell) from 10 to 30 min after mixing the reagents. However, the plot of absorbency against beryllium concentration is not linear; the curvature is considerable at concentrations above 16 µg Be per 50 ml.

References p. 131

The berillon method has been adapted to field conditions to provide a rapid and sensitive beryllium analytical procedure for use in geochemical prospecting[14]. The colours obtained are compared visually with those of standards; a maximum sensitivity of 0.5 p.p.m. Be can be achieved by an experienced operator.

The most sensitive chemical test known for beryllium is that based on the fluorescence developed when beryllium reacts with 3,5,7,2',4'-pentahydroxyflavone (morin) in sodium hydroxide solution of pH greater than 11.0[15]. Careful control of conditions (temperature, the reagent concentrations) is said to give a detection limit of 0.004 μg Be and a precision of 0.8% on 0.2 μg Be at the 95% confidence level. However, the method is not specific for beryllium and is subject to many interferences, so that beryllium must first be separated from the other elements present[16, 17].

An important application of the morin fluorescence method is for the routine determination of microgram or submicrogram amounts of beryllium in filter paper, such samples arise in large numbers in beryllium laboratories from the operation of health monitoring services (see Chapter 10). Separation of beryllium from bulk impurities (such as might be present in a smear sample) is carried out by chelating the latter with sodium diethyl dithiocarbamate (which does not complex beryllium) and then extracting the thiocarbamate complexes with ethyl acetate[18, 19]. Trace impurities (such as would be present in airbourne dust samples) are effectively sequestered with E.D.T.A. Iron, which is not complexed by E.D.T.A. at the high pH values used in the morin method, is sequestered by citric acid and triethanolamine.

It should be noted that the purity of the morin reagent is of importance, even the best quality commercial material always gives high blank readings. These blanks are reduced by one half by dissolving the morin in alcohol, standing the solution over activated charcoal and reprecipitating the morin by the addition of ether[18].

Another very sensitive reagent for beryllium is Chrome Azurol S, the sodium salt of 3''-sulpho-2'',6'-dichloro-3,3'-dimethyl-4-hydroxy-fuchson-5,-5'-dicarboxylic acid (British Colour Index No 723). This compound gives a pink to purple blue colour with beryllium in neutral or weakly acidic solutions[20, 21]. Silverman and Shidler[22] have shown that for optimum results the solution should be buffered to pH 6.0 ± 0.1 by means of pyridine-hydrochloric acid, since the intensity of colour varies with the pH. The time for reading is not critical, but consistent results are assured when the liquid is allowed to stand for fifteen minutes before measuring the colour intensity. Unfortunately, the method is susceptible to interferences, positive by metals such as Fe^{3+}, Al^{3+}, Zr^{4+} and Pb^{2+}, negative with complexing agents such as acetate, tartrate or E.D.T.A. It appears that to obtain maximum sensitivity, and to ensure accurate results, all interfering species must be absent. Under these ideal conditions from 0.2–10 µg Be per 50 ml can be determined with a precision of 0.2 µg. Like all the metal–Chrome Azurol S complexes, that formed by beryllium is destroyed (bleached) by fluoride. Indeed, this reaction is employed as a sensitive method for fluoride analysis, from 1–30 µg.F per 50 ml can be determined with a precision of 1 µg[22].

Eriochrome Cyanine R is another sensitive, although not specific, reagent for beryllium[23]. Combined with the ring-oven technique this reagent has been used for the rapid analysis of beryllium in air monitoring samples[24]. Determinations can be made on 0.05 µg Be with an average error of $7 \pm \%$.

4. Newer methods for beryllium separation

In the procedures already outlined, selectivity was introduced by the addition of reagents, such as E.D.T.A. or 8-hydroxyquinoline, which either complex or precipitate the impurities rather than the beryllium. Besides these, there are a number of separation procedures which utilize the newer separation techniques. Thus beryllium can be separated from aluminium or iron by passing a solution of these metals at pH 3.5, and containing E.D.T.A., through a column of a sulphonic acid cation-exchange resin. The uncomplexed Be^{2+} ions are absorbed by the resin, whilst the anionic or neutral iron and aluminium complexes pass on[25]. Rather better results are obtained with a phosphonic acid or a diallylphosphate polymer resin[26], as these possess a much greater affinity for the Be^{2+} ions (see p. 17). Alternatively, beryllium can be separated from aluminium, iron and other trivalent metals by selective elution from a sulphonic acid resin with 1.0 N hydrochloric or 1.2 N nitric acids, the beryllium being eluted in a band ahead of the trivalent metals[27].

Solvent extraction procedures involving direct beryllium extraction have mainly used acetylacetone. Typical is the method of Adam, Booth and Strickland[28] in which beryllium is separated from other elements by a double extraction of beryllium acetylacetonate into chloroform from aqueous media at pH 7–8. The addition of E.D.T.A. to the aqueous solution before each extraction avoids interferences. The light absorption of the beryllium complex is measured in the final chloroform extract at 295 mμ against a chloroform blank. Between extractions the chloroform layer is evaporated, organic matter destroyed by wet oxidation with nitric or perchloric acids, and the residue redissolved in water. The method suffers from being manipulatively tedious, moreover, unaccountable variations have been found in the calibration graphs plotted at different times[7].

Paper chromatographic methods have also been employed. Beryllium is separated from aluminium by downward elution with a solvent consisting of 80% (by volume) methylethyl ketone and 20% hydrochloric acid[7]. The beryllium band is detected by spraying with morin or alizarin solution; the particular strip is excised, ashed, and the beryllium is dissolved in dilute sulphuric acid and determined absorptiometrically.

An improved procedure has been recently described[29] in which the chromatogram is developed on paper impregnated with the disodium salt of E.D.T.A., methylethyl ketone–hydrochloric acid or ethanol–hydrochloric acid–water is used as the developing solvent and Eriochrome cyanine R as the colour forming reagent. The blue-violet beryllium–dye complex is compared with standard chromatograms for quantitative work and as little as 0.002 μg Be can be visually detected. Impregnation with E.D.T.A. reduces interference from metals which form coloured complexes with Eriochrome cyanine R. It does however cause the beryllium band to be somewhat diffuse and trailing, a difficulty which is overcome by allowing the band to migrate only a short distance (4 cm). The sheet is finally air dried, exposed to ammonia gas and then sprayed with 0.075% Eriochrome Cyanine R solution.

5. Radiochemical separation procedures

These are dealt with separately from the macro methods because the active components in most of the solutions encountered in radiochemistry are in extremely dilute solution. Aspects of beryllium chemistry which are of little importance at macro concentrations become significant at high dilutions. For example, the ready hydrolysis of the Be^{2+} ion (see p. 8-11) leads to the adsorption of beryllium on to the walls of the containing vessel, presumably as a polymeric hydrolytic species. It has been found that up to 20% of the ^7Be from a carrier free solution in 0.1 M sodium chloride buffered with 0.001 M sodium acetate is adsorbed on to the surface of a glass container at pH 6, whereas less than 5% of the ^7Be is adsorbed on polyethylene under similar conditions. The adsorption of beryllium on to polyethylene or glass falls to zero at about pH 5 and 4.5, respectively[30]. Thus, for storage of any very dilute beryllium solution, polyethylene vessels should be employed and the pH kept below 4.

Another important consequence of the hydrolysis of the Be^{2+} ion is its ready co-precipitation with almost any precipitate formed at pH $\sim$ 7. Gelatinous hydroxides, such as those of aluminium or iron, are particularly effective. From ferric hydroxide used as a co-precipitant, beryllium can be recovered by treatment with cold caustic soda.

Solvent extraction procedures are of importance in connection with beryllium radiochemistry, those involving acetylacetone[28, 31], thenoyltrifluoro–acetone (TTA)[32], and the chloroform extraction of beryllium oxide acetate are typical. Extraction by TTA (0.01 M

References p. 131

TTA in benzene) differs from that by acetylacetone in being slow; at the optimum aqueous phase pH of 7, it takes about three hours to complete. Aluminium is also removed but less iron at pH 7 than at a lower pH. Back-extraction of the organic phase with concentrated hydrochloric acid removes all the calcium, iron, strontium and yttrium in 15 minutes. These are completely separated from the beryllium; this requires at least 80 hours (*cf.* 6 h for aluminium) for complete back–extraction. Beryllium is however rapidly and completely stripped from the organic phase by a mixture of two parts concentrated formic to one part concentrated hydrochloric acid. For work at tracer concentrations, the TTA method has an advantage over the acetylacetone procedure since loss of beryllium does not occur through volatilization of the beryllium–TTA complex. Separations based on the extraction of the oxide acetate have the advantage of being nearly specific for beryllium. However, the preparation of beryllium oxide acetate is relatively lengthy, and this tends to limit the utilization of the method.

Chromatographic ion-exchange methods have been applied to separations of radiochemical beryllium. For example, 1.5 M hydrochloric acid has been used for the separation of the alkaline earth elements absorbed on the sulphonic acid resin Dowex 50[33]. The metals are eluted in the expected order (see p. 16), namely, beryllium first, followed in order by magnesium, calcium, strontium and barium. Other selective eluting agents have been employed with sulphonic resins: 0.55 M ammonium lactate at pH 5 for separating beryllium and the alkaline earths[33], 0.02–0.10 M sulphosalicylic acid[34] at pH 3.5–4.5 for separating beryllium from Cu^{2+}, UO_2^{2+} and Ca^{2+}, and dilute magnesium or calcium solutions for separating beryllium and aluminium[35]. The beryllium is always eluted first, rapidly with >3 M hydrochloric acid[25, 36], or by 0.5 M sodium acetate in 1.0 M acetic acid[30]. With the latter all the beryllium is removed in only 0.5–1.5 column volumes of effluent.

6. Radiometric methods of beryllium analysis

Following the pioneering work of Gaudin[37], several instruments have been devised for the radiometric assay of beryllium, based upon the nuclear reaction:

$$^9_4Be + h\nu = {}^8_4Be + {}^1_0n$$

In this the number of neutrons produced is directly proportional to the quantity of beryllium present. Provided that the source and the geometry of the system remain unchanged, the beryllium contents of different samples can be compared.

An example of such an instrument is the Beryllium Assay Equipment type 1646A designed and constructed by the U.K. Atomic Energy Authority[38]. The γ source is an aluminium capsule containing antimony; this has been pile irradiated to give ^{124}Sb (60 day half life) and ^{122}Sb (3 day half life). It is the 2.39 MeV γ radiation from the ^{124}Sb which liberates neutrons from the beryllium atoms (see p. 144), the ^{122}Sb merely causes an increase in the background gamma radiation. The source is usually a 100–125 millicuries (mC) of ^{124}Sb and is used for about six weeks before re-irradiation[39]. The neutrons liberated are counted by five boron trifluoride counters appropriately disposed around the sample. Indeed, the maintenance of correct counter geometry at all times is one of the most important features in the design of this equipment.

The sensitivity and accuracy of the γ,n assay equipment is enhanced by an increase in source size: the latter because the statistical accuracy of a determination depends upon the number of counts. For instance, at least 6000 total counts are required for a 2% accuracy. Obviously with a particular beryllium content this minimum number of counts can be achieved more rapidly with a larger gamma source. Alternatively, the amount of beryllium required to give sufficient counts in a given time decreases as the intensity of the source is raised; that is the sensitivity is increased. But large sources present shielding problems, and this limits the power of the source that it is convenient to employ.

An advantage of the γ,n method is that it provides non-destructive analysis of solid samples. This both by-passes the time-consuming process of sample dissolution, and allows valuable experimental material to be preserved. Samples of about 50 g are usually employed in the 1646A equipment, although simple modifications allow the use of smaller samples. The count rate for the unknown sample is compared with that for a standard under similar conditions, both being corrected for background. The percentage of beryllium in the sample is given by:

$$\frac{\text{counts per min for sample/weight of sample}}{\text{counts per min for standard/weight of standard}} \times \% \text{ Be in standard}$$

For accuracy, the volume of sample and standard should be about the same, and their densities and particle sizes should be similar.

The procedure is applicable to beryllium containing solutions when the count rates of equal volumes of sample and standard are compared. It provides the only analytical method for beryllium solutions with which fluoride does not interfere. With an ^{124}Sb source of about 100 mC, the background count is about 4 per minute. The count rate for a 50 ml liquid sample containing 1 mg BeO per ml is about 120 counts/min; this is about 2 counts/min per mg BeO. Similar figures apply to solid samples.

The only elements which interfere with the γ,n method are the rare earths (particularly gadolinium) and boron[38] which absorb neutrons and reduce the observed count rate. However, these elements must be present in considerable quantity for their interference to be significant. Although the γ,n method is invaluable in a laboratory undertaking a large number of beryllium analyses, it uses expensive equipment and has a large space requirement. However, within its limitations, it probably represents one of the most perfect analytical procedures ever devised.

The γ,n method has been adapted for the detection and assaying of beryllium minerals under field conditions[40]. The equipment is handled by means of long hinged arms which enable the operative portion to be brought into contact with any rock surface, while the operators remain at a safe distance from the gamma source. It provides an extremely powerful tool in beryllium prospecting.

7. Spectrometric methods

In general, the determination of beryllium is better undertaken either by wet analytical methods, or by the γ,n procedure, than by means of emission spectrometry. Advantages of the latter are its specificity, high sensitivity and, especially, its ability to cope rapidly with large numbers of samples. The method is empirical and a range of standard mixtures for each type of sample to be analysed must be prepared. Although a number of spectrometric procedures have been described for the determination of beryllium in beryl[41-43], the method is chiefly used for the rapid control of mineral separations and for sorting large numbers of ore samples. Even here the γ,n method is preferred if 10–100 g samples are available, containing not less than 0.01% Be (these limits are for a 100 mC source).

Emission spectrometry is very useful for the analysis of trace quantities of beryllium, especially samples produced by environmental monitoring[1, 44]. An especially important application is the development of portable spectrographic equipment for the continuous analysis of beryllium in air. Churchill and Gillieson[45] introduced the technique of continuously drawing a constant flow of air across a spark gap. The light from the spark was dispersed by a grating monochromator set for the beryllium doublet at 3130 Å and the result continuously recorded by means of a photomultiplier, d.c. amplifier and chart recorder. An improved version of this monitor has been described[46] in which integration over a short time is used rather than continuous recording. It is claimed that by careful design of spark gap and by using a pulsed-arc discharge axially in the air stream, rather than a condensed spark across it, the results are independent of the particle size and chemical composition of the airborne beryllium dust.

An important use of emission spectrometry is in the determination of impurities in beryllium oxide, beryllium metal etc. The published work in this field is very considerable (cf. ref. 1), and only one example method will be described. In it[47] the sample (2 g) is converted to beryllium oxide acetate, which is dissolved in chloroform and the impurities re-extracted into 4 N hydrochloric acid together with about 5% of the beryllium. This beryllium is reconverted to oxide, and contains all the impurities present in the original sample. These are determined by arcing the oxide in a graphite cup at 12 A, and the detection limits are stated to be:

    5   p.p.m. for Zn
    3   p.p.m. for Ca and Al
    2   p.p.m. for Ba, Ti, Fe, Sb, Te, In and Tl
    1.0 p.p.m. for Mg, Mo, Co, Ni, Sn, Pb and Na
    0.5 p.p.m. for V, Cr, Bi and Ga
    0.3 p.p.m. for Cu
    0.2 p.p.m. for Ag
    0.1 p.p.m. for Mn
    0.05 p.p.m. for Cd

8. *Polarographic methods*

Heyrovsky and Berezicky[48] did not observe a reduction wave with solutions of pure beryllium salts before the onset of the hydrogen wave caused by the hydrolysis of the Be^{2+} ion. Later, solutions of beryllium sulphate and chloride were shown to give two waves, the first at about -1.4 V due to the discharge of hydrogen and the second at -1.8 V *vs.* the S.C.E. due to reduction of the Be^{2+} ion[49], but the beryllium wave was not well defined. More recently it has been reported that an analytically usable wave is produced in an 0.5 M lithium chloride supporting electrolyte of pH 3.4, the diffusion current remaining proportional to the beryllium concentration up to 8×10^{-3} M[50]. The beryllium wave is close to that of hydrogen, so that the pH must be carefully controlled in order that the height of the beryllium wave may be measured. It does not appear at present that direct polarographic reduction methods hold much promise for beryllium analysis.

REFERENCES

1 SMYTHE, L. E. AND WHITTEM, R. N., *Analyst*, 86 (1961) 83; see also *The Determination of Beryllium*, H. M. Stationery Office, London, 1963.
2 SMITH, G. H., *Unpublished observation*, The National Chemical Laboratory.
3 VINCI, F. A., in *Standard Methods of Chemical Analysis*, Ed. by FURMAN, H. H., 6th ed., D. van Nostrand, New York, 1962; RODDEN, C. J. AND VINCI, F. A., Chapter 12, in *The Metal Beryllium*, Ed. by WHITE, D. W. AND BURKE, K. E., The American Society for Metals, Cleveland, 1955.
4 BREWER, P. J., *Analyst*, 77 (1952) 539.
5 ROGERS, N. E. AND PRATHER, D. W., *Anal. Chem.*, 31 (1959) 1081.
6 HURE, J., KREMER, M. AND LE BERQUIER, F., *Anal. Chim. Acta*, 7 (1952) 37.
7 HUNT, E. AND MARTIN, J. V., *Paper given at the symposium on the analytical chemistry of beryllium, Blackpool*, 1960. *U.K.A.E.A. report P.G. 171*. Ed. by Metcalfe, J. and Ryan, J.
8 PIRTEA, J. AND CONSTANTINESCU, V., *Z. Anal. Chem.*, 165 (1959) 183.
9 MISUMI, S. AND TAKETATSU, T., *Bull. Chem. Soc. Japan*, 32 (1959). 593.
10 MELICK, E. S., *U.S.A.E.C. report TID-7555*, 1958.
11 WALLACE, C. G., *U.K.A.E.A. reports A.E.R.E. AM 29 and AM 30*, 1959.
12 STROSS, W. AND OSBORN, G. H., *J. Soc. Chem. Ind.*, 63 (1944) 249; *Metallurgia*, 30 (1944) 417.
13 KARANOVICH, G. C., *Zhur. Anal. Khim.*, 11 (1956) 400.
14 HUNT, E., STANTON, R. E. AND WELLS, R. A., *Bull. Inst. Min. Metall.*, 69 (1959-60) 361.
15 TORIBARA, T. Y. AND SHERMAN, R. E., *Anal. Chem.*, 25 (1953) 1594.
16 SILL, C. W. AND WILLIS, C. P., *ibid.*, 31 (1959) 598; 33 (1961) 1671.
17 FLORENCE, T. M., *Anal. Chim. Acta*, 20 (1959) 472.
18 BERKLEY, A. M. AND SMITH, G. H., *Scientific Report NCL/AE 203*. The National Chemical Laboratory, 1962.
19 VANOSSI, R., *An. Asos. Quim. Argentina*, 45 (1957) 215.
20 THEIS, M., *Z. Anal. Chem.*, 144 (1955) 192.
21 WOOD, J. H., *Mikrochim. Acta*, (1955) 11.
22 SILVERMAN, L. AND SHIDLER, M. E., *U.S.A.E.C. report NAA-SR-2686* 1958.
23 HILL, U. T., *Anal. Chem.*, 30 (1958) 521.
24 WEST, P. W. AND MOHILNER, R. R., *ibid.*, 34 (1962) 558.
25 NADKARNI, M. N., VARDE, M. S. AND ATHERVALE, V. T., *Anal. Chim. Acta*, 16 (1957) 421.
26 KENNEDY, J. AND WHEELER, V. J., *ibid.*, 20 (1959) 412.
27 STRELOW, F. W., *Anal. Chem.*, 33 (1961) 543.
28 ADAM, J. A., BOOTH, E. AND STRICKLAND, J. D. H., *ibid.*, 6 (1952) 462; see also ALIMARIN, I. P. AND GIBALO, I. M., *Zhur. Anal. Khim.*, 11 (1956) 389.
29 ADER, D. AND ALON, A., *Analyst*, 80 (1961) 125.
30 FAIRHALL, A. W., *U.S.A.E.C. report NAS-NS-3013*, 1961.
31 TORIBARA, T. Y. AND CHEN, P. S., *Anal. Chem.*, 24 (1952) 539.
32 BOLOMEY, R. A. AND WISH, L., *J. Amer. Chem. Soc.*, 72 (1950) 4483.

33 MILTON, G. M. AND GRUMMITT, W. E., *Can. J. Chem.*, 35 (1957) 541.
34 SCHUBERT, J., LINDENBAUM, A. AND WESTFALL, W., *J. Phys. Chem.*, 62 (1958) 390.
35 HONDA, M., *J. Chem. Soc. Japan*, 71 (1950) 118; *ibid.*, 72 (1951) 361; KAKIHANA, H., *ibid.*, 72 (1951) 203.
36 DIAMOND, R. M., *J. Amer. Chem. Soc.*, 77 (1955) 2978.
37 GAUDIN, A. M. AND PANNELL, J. H., *Anal. Chem.*, 23 (1951) 1261.
38 MILNER, G. W. AND EDWARDS, J. W., *U.K.A.E.A. report AERE R2965* (1959); MILNER, G. W. *et al.*, *ibid.*, *AERE R3212* (1960); BISBY, H. *ibid.*, *AERE R3021* (1959); *Nuclear Power*, 5 (1960) 100.
39 AUDRIC, B. N., *unpublished observations*, The National Chemical Laboratory.
40 BOWIE, S. H. U. *et al.*, *Bull. Inst. Min. Metall.*, 69 (1959-60) 345.
41 KEHRES, P. W. AND POEHLMANN, W. J., *J. Appl. Spectroscopy*, 8 (1954) 36.
42 ALEKSEEVA, V. M. AND RUSANOV, A. K., *Zhur. Anal. Khim.*, 12 (1957) 23.
43 CREITZ, E. C., *U.S. Bureau of Mines Report No. 5407*, Washington, D. C. (1958).
44 BROOKS, R. O., *Nuclear Power*, 3 (1958) 112 and 539.
45 CHURCHILL, W. L. AND GILLIESON, A. H., *Spectrochim. Acta*, 5 (1952) 238.
46 WEBB, R. J., WEBB, M. S. AND WILDY, P. C., *U.K.A.E.A. report AERE R2868* (1959).
47 KARABASH, A. G. *et al.*, *Zhur. Anal. Khim.*, 14 (1959) 94.
48 HEYROVSKY, J. AND BEREZICKY, S., *Coll. Czech. Chem. Comm.*, 1 (1929) 19.
49 KEMULA, W. AND MICHALSKI, M., *ibid.*, 5 (1933) 436.
50 VENKATARATNAM, G. AND RAGHAVA RAO, B. S., *J. Sci. Ind. Res. India*, 17B (1958) 436; see also GYORBINO, K., *Acta Chim. Hung.*, 22 (1960) 225.

CHAPTER 10

The Beryllium Health Hazard and its Control

Here is given an account of the precautions considered necessary for the safe operation of a beryllium laboratory. The number of cases of beryllium disease now well authenticicated show that the hazards involved in working with beryllium and its compounds must not be underrated. Nevertheless, when the safety rules are carefully followed the risk is infinitesimal. It must be emphasised that the present account simply provides an outline of the nature of the problem and of the kind of precautions which have to be taken. Usually laboratories handling beryllium have been designed under expert guidance and the personnel are subject to regular medical examination.

1. An outline of the occurrence and nature of beryllium disease

The report of Weber and Engelhardt in 1933[1] describes the first occasion on which the incidence of disturbing respiratory conditions, bronchitis and bronchiolitis, was observed in beryllium workers. Soon afterwards Gelman[2], and others of the Obuch institute for Occupational Diseases in Moscow, drew attention to the high incidence of respiratory diseases of men employed in the extraction of beryllium from its ores. Gelman recommended that work on the extraction of beryllium be undertaken only in a room well ventilated at the point where dust or fumes are produced. The workers should observe high standards of personal hygiene; gloves and other protective clothing should be worn and washing accommodation, douche baths etc. be provided.

The hazardous nature of beryllium compounds became more

References p. 140

generally appreciated during the second World War when cases in Germany, Italy, Russia and the United States were reported. The trouble was mainly with those in beryllium extraction plants, although workers in the fluorescent lamp industry which then used zinc beryllium manganese silicate were also affected. Gradually evidence from a diversity of sources placed beyond question the toxic character of beryllium and its compounds. These reports disclosed an alarming and characteristic feature of the disease, namely the long time which can elapse between exposure to beryllium compounds and the onset of symptoms. Delay periods of up to ten years have been reported by Hardy[3].

The Beryllium Register of the Massachusetts General Hospital (opened in 1952) had recorded 616 patients by January 1st, 1960, the overall mortality rate being about 20%. However, the adoption of suitable precautions has reduced the number of cases reported; there were only ten between 1949–1958 in which exposure had occurred after 1949, the date when standards of operation were adopted. There have been only two fatal cases in Great Britain, and both occurred before 1949.

The chief hazard of working with beryllium or its compounds is the inhalation of the material as dust, smoke or aerosols. All simple beryllium compounds appear to be dangerously active: BeF_2, BeO, $Be(OH)_2$, $BeSO_4$, $BeCl_2$ etc. The mineral beryl ($3BeO.Al_2O_3.6SiO_2$) is not considered a hazardous material; no information appears to be available concerning the toxicity of other beryllium minerals, although the toxic character of the beryllium silicate phosphors (these are related to phenacite) should be noted.

Inhalation of these solids leads to pneumonitis, a chronic disease of the lung tissue. This, in either its acute or chronic form, constitutes the chief health hazard associated with beryllium. In one well documented case[4], in which the beryllium fluoride concentration in the air rose to about 450–600 μg Be/m^3, two out of eight workers developed bronchitis and one pneumonitis (the acute form of beryllium disease). Unlike his fellows, the latter worker had been undertaking manual work at the time of exposure, and his resulting higher rate of breathing must have caused greater intake of toxic material.

1 OCCURRENCE AND NATURE OF BERYLLIUM DISEASE

Beryllium disease had also been diagnosed amongst people living near beryllium plants, these being designated "neighbourhood cases". Forty seven well established neighbourhood cases have now been recorded. An extreme example was that of the housewife who is believed to have contacted the disease through washing the clothes of her husband, a beryllium worker. It has been estimated that an average of 17 μg Be could be inhaled by a person laundering such contaminated overalls.

Contact dermatitis has been reported as arising from exposure to beryllium containing dusts and fumes, sometimes accompanied by conjunctivitis. Penetration of cuts and abrasions by beryllium compounds may also lead to the production of ulcers which persist until the beryllium is removed. It is thus necessary for care to be taken to prevent beryllium compounds, either in the solid state or in solution from coming into contact with the skin.

For a more detailed discussion and bibliography of the occurrence and symptoms of beryllium disease see refs. 5–7.

2. Health control in a beryllium laboratory

A laboratory worker handling beryllium containing materials must be *continually aware of the potential danger, in particular the inhallation of beryllium containing dusts*. The utmost cleanliness in all manipulation is of prime importance and all experiments must be so designed as to eliminate the possibility of release of beryllium containing dusts into the atmosphere. It cannot be over-emphasised that good housekeeping in the laboratory is one of the most important aspects of this safe handling. There is far less risk of accident when the benches are kept free of unnecessary equipment, and all apparatus cleaned immediately after use and stored in a closed cupboard. The benches, including bench trays, should be cleaned at the end of each working day; and, at the end of every week *all* benches, fume cupboards, reagent bottles, writing desks etc. should be washed down and all apparatus washed and put away. These cleaning operations should form a rigid routine.

A high degree of personal hygiene must be observed by all workers in a beryllium area. Eating and smoking in the area should be absolutely prohibited and protective clothing worn at all times; these should be laundered at regular intervals. Workers must always remove their protective clothing and wash before leaving a beryllium area. Under normal conditions respirators are not necessary, but for some specially hazardous operations, and always where there is any doubt, they should be worn.

As the dangers arise mainly from dust, it is essential that any operation which could possibly generate beryllium dust must be done in a glove box or adequately ventilated fume cupboard; for instance, the ignition of gravimetric precipitates, or the handling of volatile compounds and dry solids. The transport of beryllium compounds about a laboratory (always to be kept to a minimum) must only be undertaken in closed unbreakable boxes, not glass containers which may break and release beryllium into the air. *At no time* must dry beryllium compounds be exposed to the open laboratory atmosphere.

It is also necessary to control the routine cleaning operations in a beryllium laboratory. Dry sweeping of floors and ledges may raise dangerous quantities of contaminated dust, and vacuum cleaners must be used. The floors, which should preferably be covered with a non-porous floor covering such as linoleum, must be polished regularly with a water-repellant wax. This reduces the chance of absorption of beryllium compounds should these be spilt.

The handling of beryllium containing solutions is much less hazardous than dry compounds, and, with care, operations with these can be performed in the open laboratory. Nevertheless, spillage of a beryllium solution creates, on evaporation, toxic beryllium dusts. The effect of spillage can be minimised by working on a non-absorbant bench tray made for instance of polyvinylchloride. Any liquid can be immediately mopped up by a damp cloth and the tray and cloth then washed. Wooden surfaces, such as bench tops, are undesirable since being porous they retain small quantities of solution which on drying creates dust.

Design is important in any beryllium laboratory. In particular, the general ventilation of the building must be highly effecient, and the extracted air must be filtered and discharged through a stack positioned so that the extracted air cannot re-enter the building. Special air extracts, fitted directly to the working enclosures such as glove boxes or fume cupboards, should reduce the pressure below that in the open laboratory. Similarly, the beryllium laboratory is itself kept at a lower pressure than the air outside. The air velocity through the openings into the working enclosures should be about 150–200 cubic feet per minute. The standard of ventilation should be sufficiently good to allow non-opening windows to be installed. The laboratory should have a minimum of dust-harbouring shelves and crevices, be well illuminated and be kept freshly painted. Fume cupboard space must be generous, and the interior should be easily and effectively washed down. The laboratory should be subdivided into at least two or three rooms; this makes it possible to reduce beryllium exposure by separating the different kinds of work.

The storage of beryllium compounds must be strictly regulated. Not more than 500 g Be total, in *any chemical form*, should be kept in a laboratory; otherwise, in the event of fire, beryllium compounds could be vaporised or blown about to such an extent as to constitute a hazard to the surrounding area. Beryllium in excess of this quantity should be separately stored in a concrete lined hole safe from fire risk. Special care must be taken in the storage and handling of volatile or spontaneously inflammable compounds, such as the beryllium alkyls.

It is considered advisable (*e.g.* ref. 5, p. 75) to select personnel for a beryllium project on the basis of medical history, a physical examination and a chest X-ray. It is usual to reject those who have a greater than average chance of developing beryllium disease through their suffering from chronic respiratory diseases or from skin diseases which might make them susceptible to beryllium dermatitis. Rejection is also best for those with abnormal chest X-rays, a history of asthma, heart disease or tuberculosis, all conditions which show symptoms that could be medically confused with changes induced by chronic beryllium disease. All workers should have a medical examination at least once a year which should include a blood count and a full chest X-ray. Careful records should be kept of the causes of all sick leave as this can provide clues as to the possible commencement of the disease.

Beryllium areas should be labled restricted zones and only medically cleared personnel allowed to enter. Indeed, it is desirable to reduce all entry to the minimum as this makes easier the rigid application of the safety code. It should be noted that maintenance work in a beryllium area raises problems, especially when this is concerned with the ventilation system. Full safety regulations must be adopted to all maintenance work connected with the running of a beryllium area.

3. Maximum permissible concentrations

Maximum permissible levels of beryllium air contamination have been suggested by the U.S. Atomic Energy Commission and by the American Industrial Hygiene Association (see ref. 7).

(a) The maximum concentration of beryllium in the laboratory air should not exceed 2 μg Be/m^3 as an average over an eight hour day.

(b) Even although the daily average may be less than 2 μg/m^3 no person should ever be exposed to concentrations greater than 25 μg/m^3 even for very short times.

(c) In the neighbourhood of any plant handling beryllium compounds the average monthly concentration at the breathing zone level should not exceed 0.01 μg/m^3. As coals can contain beryllium, sometimes in appreciable concentrations (coal ashes containing 0.1–1.0% Be have been reported)[8], a substantial background beryllium concentration can occur near coal burning plants.

The figure of 2 μg/m^3 maximum average air concentration has no sound scientific basis[5] and is probably highly conservative. However, experience has shown that with a carefully run laboratory, which has efficient ventilation, and in which the maximum quantity of beryllium in use at any one time was less than 500 g, the average air contamination is normally well below 2 μg/m^3. Over 90% of the determinations over a long period showed the beryllium air concentration [9] to be between 0–0.2 $\mu g/m^3$.

The maximum beryllium concentration of 25 $\mu g/m^3$ is based on rather better scientific evidence, as it is related to both human cases and to animal experiments. It is difficult to measure instantaneous air concentrations of this kind, although portable continuously operating spectrographic equipment has been designed for this purpose (see p. 129). Unfortunately, such equipment is bulky, expensive and not yet completely reliable, so that it is not universally employed. It is emphasised that especial precautions must be taken when commencing any new laboratory operation with beryllium to ensure that at no time can the 25 $\mu g/m^3$ level be exceeded. At the start, continual air-sampling should be undertaken and, if any doubt exists, the operators must wear respirators until the safety of the operation has been proved.

Air-sampling is usually undertaken by passing a given volume of air through a filter medium such as filter paper, *e.g.* Whatman 41. A special air-sampling unit must be used for this purpose (*e.g.* ref. 10), which are in principle only modified vacuum cleaners. The paper is then ashed and analysed for beryllium (see p. 122). The weak point of this technique is that, whatever filter medium is employed, there will be some particles too small to be retained by it and these smaller particles probably constitute the greatest hazard, *e.g.* the results of Hall *et al.*[11] on the toxic effect of beryllium oxide produced under different conditions.

In addition to air samples, smear tests are taken from a given area of bench or wall by wiping it over with tissues, which are then ashed and analysed. Such measurements show if any undetected contamination of the laboratory fabric has occurred. They are also useful for detecting operations, or even operators, responsible for causing beryllium contamination.

It should be noted that active and inactive forms of beryllium (*e.g.* beryl) cannot be distinguished in the above environmental monitoring tests. Thus full precautions have to be applied in a beryllium laboratory even when handling inactive beryl, as the air-monitoring tests are invalidated by the presence of beryl dust in the atmosphere.

References p. 140

REFERENCES

1 WEBER, H. H. AND ENGELHARDT, W. E., *Zentr. Gewerbehyg. Unfallverhüt.*, 10 (1933) 41.
2 GELMAN, I., *J. Ind. Hyg. Toxicol.*, 18 (1936) 371.
3 HARDY, H. L., *A. M. A. Arch. Ind. Health*, 11 (1955) 273.
4 EISENBUD, M., et al., *J. Ind. Hyg. Toxicol.*, 31 (1949) 282.
5 TEPPER, L. B., HARDY, H. L. AND CHAMBERLIN, R. I., *Toxicity of Beryllium Compounds*. Elsevier, Amsterdam, 1961.
6 HUTCHINSON, E. D., et al., *U.S.A.E.C. report UR-570*, 1960.
7 BRESLIN, A. J. AND HARRIS, W. B., *U.S.A.E.C. report HASL-36*, 1958.
8 GOLDSCHMIDT, V. M. AND PETERS, C. I., *Nachr. Ges. Wiss. Göttingen*, 4 (1933) 371.
9 *Unpublished observations*, The National Chemical Laboratory.
10 EISENBUD, M., Chapter XII in *The Metal Beryllium*, Ed. by WHITE, D. W. AND BURKE, J. E., The American Society for Metals, Cleveland, 1955.
11 HALL, G. R., et al., *Arch. Ind. Hyg. Occup. Med.*, 2 (1950) 25.

CHAPTER 11

Nuclear Properties and Reactions of Beryllium

Beryllium has only one stable isotope ^{9}Be, the most important nuclear property of which is its very low absorption cross-section with respect to thermal neutrons. It is computed that the amount of ^{10}Be produced by three years irradiation of 1 g beryllium-9 at a flux of 4×10^{14} thermal neutrons per sq. cm. per second would afford an activity of only six microcuries. Coupled with its low mass, which allows efficient transfer of momentum with colliding particles, its low thermal neutron cross-section makes beryllium an excellent neutron moderator for nuclear reactors and a good neutron reflector for preventing the escape of neutrons from a reactor core. It has also been proposed as a canning material for the uranium fuel elements in the Advanced Gas Cooled Reactor.

In addition to the stable isotope beryllium-9, the element has four radioactive isotopes with masses six, seven, eight and ten; their properties are listed in Table 6[1].

The most commonly encountered of these active isotopes is ^{7}Be which is frequently employed as a beryllium tracer by counting its gamma radiation. It is prepared by the proton bombardment of ^{12}C or ^{7}Li in a cyclotron:

$$^{12}_{6}\text{C} + ^{1}_{1}\text{H} \rightarrow ^{7}_{4}\text{Be} + 3^{1}_{1}\text{H} + 3^{1}_{0}\text{n}$$

$$^{7}_{3}\text{Li} + ^{1}_{1}\text{H} \rightarrow ^{7}_{4}\text{Be} + ^{1}_{0}\text{n}$$

The ^{10}Be isotope is prepared by the interaction of an alpha particle with ^{7}Li:

$$^{7}_{3}\text{Li} + ^{4}_{2}\text{He} \rightarrow ^{10}_{4}\text{Be} + ^{1}_{1}\text{H}$$

This method is low-yielding and difficult and makes ^{10}Be expensive

References p. 145

to prepare. The n,γ reaction commonly employed for the preparation of active isotopes one unit greater in mass than the stable isotope is, as we have seen, inapplicable to this element. Beryllium-10 is the longest lived of the active isotopes and, as it is also a beta emitter, it would be a convenient isotope for tracer studies.

TABLE 6

RADIOACTIVE ISOTOPES OF BERYLLIUM

Isotope	Mass on physical scale	Half-life	Decay product and radiation
$^{6}_{4}Be$	6.023	approx. 0.4 sec	?
$^{7}_{4}Be$	7.019	53 days	$^{7}_{3}Li$ by orbital electron capture with 14% emission of γ rays
$^{8}_{4}Be$	8.007	1 x 10^{-15} sec	$^{8}_{4}Be \rightarrow 2\,^{4}_{2}He$
$^{10}_{4}Be$	10.168	2.5 x 10^6 years	$^{10}_{4}Be \rightarrow e^{-} + ^{10}_{5}B$

The ^{9}Be isotope takes part in a number of nuclear reactions with charged particles since the low charge on the beryllium nucleus makes the reaction easier than with heavy elements. An example is that with alpha particles:

$$^{9}_{4}Be + ^{4}_{2}He \rightarrow ^{12}_{6}C + ^{1}_{0}n$$

This is the reaction which enabled Chadwick[2] to discover the neutron. It is exothermic, with a Q value of 5.81 MeV, but the charge barrier of the nucleus gives a negligible reaction cross-section at alpha-particle energies below 0.5 MeV; above this figure, the cross-section increases with the alpha-particle energy. The Q value, or energy change involved in a nuclear reaction, is defined as the gain in kinetic energy of the ensemble of particles. The reaction also yields gamma rays because the ^{12}C nuclei are formed in an excited state; this reverts to the ground state by gamma ray emission (principally 4.4 MeV).

NUCLEAR PROPERTIES AND REACTIONS 143

This α,n reaction is commonly employed as a laboratory source of neutrons. The beryllium must be intimately mixed with an alpha source, such as radium, because the alpha particles have a very short range in the metal. The energy spectrum and yield of neutrons varies with the grain size of the mixture[3]. This can be overcome by employing plutonium as the alpha-source in the form of the homogeneous $PuBe_{13}$ alloy.

Neutrons are also liberated from beryllium by the p,n reaction:

$$^9_4Be + ^1_1H \rightarrow ^9_5B + ^1_0n$$

The Q value for this reaction is −1.85 MeV. This means that the reaction will not proceed until the proton and the ^9Be nucleus have a minimum energy of −1.85 MeV, or until the proton has a minimum energy of $10/9 \times 1.85 = 2.06$ MeV, the latter is the proton threshold energy for the p,n reaction to occur. This reaction is employed to obtain neutrons from an internal cyclotron target and is a good method of producing monoenergetic neutrons[4].

Three other nuclear reactions occur with protons:

$$^9_4Be + ^1_1H \rightarrow ^6_3Li + ^4_2He \quad Q = 2.13 \text{ MeV}$$

$$^9_4Be + ^1_1H \rightarrow ^8_4Be + ^2_1D \quad Q = 0.56 \text{ MeV}$$

$$^9_4Be + ^1_1H \rightarrow ^{10}_5B + h\nu \quad Q = 6.6 \text{ MeV}$$

Four reactions occur with deuterons, the most important of which is the exothermic d,n reaction:

$$^9_4Be + ^2_1D \rightarrow ^{10}_5B + ^1_0n \quad Q = 4.36 \text{ MeV}$$

This reaction is a prolific source of neutrons, the yield increasing with deuteron energy above the threshold of 0.9 MeV. Owing to the exothermic nature of the reaction, the neutrons produced have up to 4 MeV more energy than the bombarding deuterons. Some slow neutrons are also produced by d,nγ reactions with the $^{10}_5B$ nuclei formed in the main reaction.

The other deuteron reactions are:–

References p. 145

$$^9_4\text{Be} + ^2_1\text{D} \to ^1_1\text{H} + ^{10}_4\text{Be} \quad Q = 4.59 \text{ MeV}$$

$$^9_4\text{Be} + ^2_1\text{D} \to ^3_1\text{H} + ^8_4\text{Be} \quad Q = 4.59 \text{ MeV}$$

$$^9_4\text{Be} + ^2_1\text{D} \to ^7_3\text{Li} + ^4_2\text{He} \quad Q = 7.15 \text{ MeV}$$

Beryllium takes part in n,α and n,2n reactions with fast neutrons:

$$^9_4\text{Be} + ^1_0\text{n} \to ^6_2\text{He} + ^4_2\text{He} \quad Q = -0.64 \text{ MeV}$$

$$^9_4\text{Be} + ^1_0\text{n} \to ^8_4\text{Be} + 2^1_0\text{n} \quad Q = -1.66 \text{ MeV}$$

The threshold values are 0.71 and 2.7 MeV respectively, the value for the n,2n reaction is greater than the theoretical value of 1.85 MeV because the ^8_4Be nucleus appears in its 2.43 MeV level rather than its ground state[5]. Both reactions may be of significance in nuclear reactors. The n,α reaction produces helium through the discharge of the alpha particles which stay in the beryllium as interstitial gas bubbles, neutron-hungry lithium-6 and tritium, the latter being formed thus:

$$^6_3\text{Li} + ^1_0\text{n} \to ^4_2\text{He} + ^3_1\text{H}$$

Although the tritium decays to ^3_2He by beta emission (half-life 12.5 years), it is rapidly reformed by thermal neutrons:

$$^3_2\text{He} + ^1_0\text{n} \to ^3_1\text{H} + ^1_1\text{H}$$

The n,2n reaction is beneficial in a reactor as it produces a large bonus of neutrons which, being of low energy, do not easily escape from the reactor. Again the helium atoms aglomerate into bubbles and cause swelling in the metal. This constitutes one of the causes of its in-pile irradiation damage.

The release of photoneutrons from ^{9}Be on gamma irradiation is one of the most important nuclear reactions of beryllium:

$$^9_4\text{Be} + h\nu \to ^8_4\text{Be} + ^1_0\text{n}$$

The Q value of −1.66 is also the threshold value as the momentum of a photon is small. The ^{8}Be decays at once into two helium atoms with about 0.09 MeV of kinetic energy between them. The threshold value of 1.66 MeV is the lowest value for photoneutron production of any element. Examples of threshold values for other elements are: deuterium 2.23, ^{17}O 4.14, ^{13}C 4.95 and ^{6}Li 5.50 MeV. Photoneutrons are produced from ^{9}Be by a number of gamma sources such as radium, ^{88}Y and ^{206}Bi, but the most convenient is ^{124}Sb which is employed in the γ,n beryllium assay equipment (see p. 127). The energy of the neutrons produced by gamma radiation from ^{124}Sb is 0.029 MeV. Higher energy neutrons are produced by other gamma sources, for instance, ^{24}Na and ^{140}La produce neutrons of 1 and 0.7 MeV energy, respectively.

Laboratory electron accelerators provide indirect gamma sources for the γ,n reaction by means of the "bremsstrahlung" (breaking radiation) that accompanies the slowing-down of electrons. It is this effect which produces the short wave radiation classed as X-rays or gamma rays in targets bombarded by fast electrons. Very high neutron fluxes can be obtained by means of electrons from high energy accelerators (*e.g.* ref. 6); the neutron production rate during the pulse of a pulsed beam of 3.2 MeV electrons can be 2×10^{12} neutrons per second. Even higher neutron production rates are achieved with a betatron, in which electrons in a circulating beam are raised to a high velocity.

When beryllium is irradiated with very high energy gamma radiation, such as is afforded by the betatron, other nuclear reactions take place:

$$^9_4\text{Be} + h\nu \to {^1_1\text{H}} + {^8_3\text{Li}} \quad Q = -16.87 \text{ MeV}$$

$$^9_4\text{Be} + h\nu \to {^2_1\text{D}} + {^7_3\text{Li}} \quad Q = -16.7 \text{ MeV}$$

$$^9_4\text{Be} + h\nu \to {^3_1\text{H}} + {^6_3\text{Li}} \quad Q = -17.7 \text{ MeV}$$

These reactions are only of academic interest at present.

REFERENCES

1 STROMINGER, D., HOLLANDER, J. M. AND SEABORG, G. T., *Rev. Modern Phys.*, 28 (1956) 559.
2 CHADWICK, J., *Proc. Roy. Soc.*, 142 (1933) 1.
3 STEWART, L., *Phys. Rev.*, 98 (1955) 740.
4 JOHNSON, V. R., *ibid.*, 79 (1950) 187.
5 FISCHER, G. J., *ibid.*, 108 (1957) 99.
6 DUCKWORTH, J. C. AND MERRISON, A. W., *Nature*, 163 (1949) 869.

Subject Index

α particles,
  reaction with beryllium, 141, 142
  emission from beryllium, 142-144
Abundance, 102
Acetylacetone,
  colorimetric reagent for beryllium, 124
  complex with beryllium, 66 et seq.
Adsorption of beryllium on walls of containing vessels, 125
Albite, formation in Copaux process, 105
Alcoholates, 99
Alloys of beryllium with
  plutonium, 143
  tantalum and zirconium, 88
Aluminium,
  chemical similarity to beryllium, 3, 4
  fractional distillation from beryllium chloride, 113
  separation of beryllium from, 111, 126
Amide, 20
Aminoberyllates, 20
Ammonobasic compounds, 20
Ammonium fluorides,
  dissolution of metallic beryllium, 3, 114
  flux, in analysis of beryllium, 118
  purification of beryllium hydroxide by, 114
  stripping agent in solvent extraction of beryllium, 112

Amphoterism of beryllium, 3, 8, 14, 20
Antimony, 124, 127
Apatite, removal from beryl flotation concentrates, 107
Arsenates, 30
Ascorbic acid, use in determination of beryllium, 121
Azide, 33

β particles, emission from beryllium, 142
Barylite, 103
Berillon, colorimetric reagent for beryllium, 121
Bertrandite, 103
Beryl,
  acid extraction of beryllium from glasses of, 109
  chlorination of, 113
  non-toxic nature of, 134, 139
  reaction with alkalies, 110
  reaction with sodium fluorosilicate, 104 et seq.
  recovery, 103
  synthesis, 34
  structure, 103
Beryllates, 14 et seq.
Beryllia, see Oxide
Beryllides, 88
Beryllium disease,
  compounds responsible for, 134
  control of, see Health control
  neighbourhood cases, 135
Beryllium metal,
  chemical properties, 2, 3
  determination of impurities in, 129

SUBJECT INDEX

nuclear properties, 141 et seq.
preparation, 113
reaction with
   ammonium bifluoride, 3
   dimethylmercury, 91
   diphenylmercury, 97
   halogens, 57, 58
   potassamide, 20
Boiling points of
   bromide, 57
   chloride, 50
   dimethylberyllium, 92
   fluoride, 42
   iodide, 58
   metal, 3
   nitride, 85
   oxide, 87
   oxide acetate, 85
Bonds, nature in compounds of beryllium, 3, 4, 38, 50, 91
Borides, 84
Born equation, 64
Borohydride, 83, 99
Boron, interference in radiometric determination of beryllium, 128
Boron trifluoride counters, 127
Bremsstrahlung, 145
Bromide, 57

Calcium beryllate, 16
Capryl alcohol, 111
Carbide,
   preparation and properties, 85
   reaction with halogens, 50, 57, 113
Carboxylates, normal salts, 32
Charge to radius ratio,
   of Al, 4, 17
   Be, 4, 8, 16, 17, 23, 64
   Ca, 4
   Li, 4
   Mg, 4
Charge number of beryllium, 1, 6, 54
Chiolite, 104
Chloride,
   dimers, 51
   electrolysis of molten, 54, 115
   fractional distillation, 113

Friedel-Crafts catalyst, 56
polymorphic forms, 50
preparation and properties, 50, 113, 115
properties of molten, 53
reaction with
   chloroform, 56
   cyclopentadienyl sodium, 98
   Grignard reagent, 92, 95, 96, 97, 99
   lithium aluminium hydride, 82
   lithium borohydride, 83
   phenol, 100
   sodium ethoxide, 99
   similarity to dimethylberyllium, 93
structure, 51
toxicity, 134
vapour density, 2
Chloride, complexes formed with
   ammonia, 54
   chloride, 52
   diethyl ether, 55
   organic ligands, 54
Chloroberyllates, 53
Chloromercurates, 53
Chromatographic methods for determination of beryllium, 124
Chrome Azurol S, colorimetric reagent for beryllium, 122
Chrysoberyl, 13, 103, 109, 110
Colorimetric determination of beryllium, 121 et seq.
Complexes of beryllium,
   factors governing stability, 64
   nature of bonds in, 4
   steric hindrance in, 18, 66
   thermodynamics, 67
   with
      carboxylic acids, see Oxide carboxylates
      catechol, 100
      chloride, 52
      citric acid, 70
      1,3 diketones, 66 et seq.
      E.D.T.A., 17, 66, 118
      fluoride, 43 et seq.

SUBJECT INDEX

hydroxy acids, 69
oxalate, 31
phosphate, 30
phosphate esters, 17
phthalocyanine, 78
salicylic acid, 69
sulphosalicylic acid, 70
Co-ordination polymers, 76
Copaux process, 104
Cryolite, 104, 106
Cyanide, 35

Dermatitis, caused by beryllium, 135
Deuterons,
  emission from beryllium, 143, 144
  reactions with beryllium, 143, 145
Dicyclopentadienylberyllium, 98
Diethylberyllium, 95
Dimethylberyllium, 33, 82, 83, 91-95
Di-(2-ethylhexyl)phosphoric acid, 18, 19, 111-112
Di-*iso*propylberyllium, 97
Diphenylberyllium, 97
Di-*tert*-butylberyllium, 82, 97

Electrolysis,
  aqueous solutions, 6, 12
  fused salts, 6, 53, 115
Electron deficient compounds, 4, 83, 92
Electronegativity, 3, 91
Electronic structure,
  beryllium, 2
  calcium, 5
  zinc, 5
Environmental monitoring, 122, 129
Eriochrome cyanine R, colorimetric reagent for beryllium, 123, 124
Ethoxide, 99
Ethyl-beryllium hydride, 96
Ethylenediamine-tetra-acetic acid, complexes with beryllium, 17, 66, 118

Filter paper, determination of beryllium in, 122
Flotation of beryl, 103

Fluorescence,
  silicates, 134
  sulphide, 88
Fluoride,
  bonds in, 38, 49
  glasses, 40, 48
  phase diagram, 41
  polymorphic forms, 40
  preparation and properties, 38 *et seq.*
  reduction, 114
  similarity to silica, 39
  systems with
    barium fluoride, 47
    caesium fluoride, 47
    calcium fluoride, 46
    lithium fluoride, 45, 46, 47
    magnesium fluoride, 44
    sodium fluoride, 45, 46, 47
    strontium fluoride, 47
    thorium fluoride, 47
    uranium fluoride, 47
    water-ammonium fluoride, 43
    water-sodium fluoride, 44, 108
Flurimetric determination of beryllium, 122
Fluoroberyllates,
  ammonium, 39, 113
  formation in Copaux process, 105 *et seq.*
  occurrence in vapour phase, 48
  preparation from aqueous solution, 44
Fluorosilicate, thermal decomposition, 105
Fuse-quench process for opening beryl, 109

γ rays,
  emission from, 142
  estimation by irradiation with, 127
  reactions with, 127, 144, 145
Gadolinium, interference in radiometric determination of beryllium, 128

SUBJECT INDEX

Geochemical prospecting, analytical methods for beryllium, 122, 128
Gravimetric methods for determination of beryllium, 118 et seq.

Halides (see also Fluoride etc.), reaction with liquid ammonia, 20
Health control in laboratories,
   air and smear sampling, 139
   good housekeeping, 135-138
   maximum permissible concentrations, 138
   medical examination, 137
   ventilation requirements, 137
Helvine, 103
Hexamminocobaltic chloride, analytical reagent for beryllium, 120
High temperature microscopy, 109
Hydrated beryllium cation,
   extraction by phosphate esters, 17
   hydrolysis, 8, 11, 16, 23, 125
   properties, 7
   sorption by ion-exchange resins, 17, 123, 124
   sound absorption by, 11
Hydride, 82
Hydrodiethyl-beryllate, 95
Hydroxide,
   acidic reactions, 14 et seq.
   occurrence in vapour phase, 13, 87
   precipitation, 9, 12 et seq., 43, 86, 105, 111, 117-119, 125
   structure, 12
   thermal decomposition, 13, 86, 119
   toxicity, 134
8-Hydroxyquinoline, precipitation by, 117

Iodate, 33
Iodide, 20, 58
Ion-exchange,
   anion-exchange resins, 24, 30, 54
   elution of beryllium from cation-exchange, 126

   on beryllia, 48
   phosphonic acid and phosphate resins, 17, 124
   sulphonic acid resins, 10, 16, 52, 123, 126
Ionic potential, see Charge to radius ratio
Ionic radii, 3, 4, 39, 52
Ionization potential, 5, 6
Irradiation damage of beryllium, 144
Isotopes of beryllium, 142

Jones and Dole equation, 7

Localised hydrolysis, 17

Melting points of
   beryl, 109
   borohydride, 83
   bromide, 57
   bromide-ether complex, 57
   chloride, 50
   chloride-ether complex, 55
   1,3 diketones, 68
   fluoride, 40, 41
   iodide, 58
   metal, 3
   nitride, 85
   oxide, 87
   oxide acetate, 71
Methoxide, 99
Methylethyl ketone, chromatographic reagent for beryllium, 124
Minerals,
   beryllium, 103
   determination of beryllium in, 120, 127
   extraction of beryllium from, 104 et seq.
Morin, fluometric reagent for beryllium, 122
Mullite, 109

Nervanaid F, use in determination of beryllium, 121
Neutrons,
   emission from beryllium, 127, 141-144

SUBJECT INDEX

reactions with beryllium, 141, 144
sources of beryllium, 143
Nitrate, 28
Nitride, 85

Olated complexes, 14
Orbitals used by, 4
Order-producing character of beryllium cations, 7, 17
Oxalate, 30
Oxide,
  boiling point, 87
  chlorination of, 50, 115
  determination of impurities in, 129
  discovery of, 1
  dissolution in beryllium salt solutions, 8, 10
  glasses containing, 88
  heat of formation, 3
  melting point, 87
  nature of bonds in, 87
  preparation, 13, 31, 86, 109
  properties, 1, 86
  reactions in fluoride melts, 48
  reduction, 87
  toxicity, 134, 139
Oxide acetate of beryllium,
  clathrate compounds, 73
  extraction by chloroform, 71, 86, 125
  hydrolysis, 75, 76, 86
  polymeric forms, 72
  preparation and properties, 71 et seq.
  purification, 71, 86
  reaction with alcohols, 74
    ammonia and amines, 73
Oxide carboxylates, 71 et seq.
Oxide carbonate, 23
Oxide fluoroberyllate, 43, 44
Oxide nitrate, 28
Oxosalts, 23 et seq.

Partial molar entropies of bipositive cations, 7
Perchlorate, 33

Periodate, 33
Periodic table, position in, 2
Phenacite,
  attack by acids, 34
  crystal form, 103
  modified form of, 109
  synthesis, 34
Phenolates, 100
Phenylberyllium bromide, 98
Phosphate of beryllium,
  preparation and properties, 29
  complexes with, 30
  in gravimetric determination of, 118, 119
Plutonium alloy with beryllium, 143
Polarographic determination of beryllium, 130
Polynuclear beryllates, 14 et seq.
Polynuclear hydrolytic species,
  formation, 8, 23 et seq.
  sorption by cation-exchange resins, 11
  sound absorption by, 11
  solvent extraction, 19
Post-heating, 109, 110
Prospecting, field methods for determination of beryllium, 122, 128
Protons,
  emission from beryllium, 141, 144
  reaction with beryllium, 141, 143
Pyrophospate, 29, 118, 119

Q value for nuclear reactions, 142

Radioactive isotopes of beryllium, 142
Relaxation measurements on beryllium solutions, 11
Respiratory diseases due to beryllium, 133

Selenate, 27
Selenide, 88
Separation of beryllium from other elements, 117 et seq., 125
Silicates, 34

SUBJECT INDEX

Silicon tetrafluoride, role in Copaux process, 105, 108
Spectrometry, emission methods for estimation of, 129
Spiran type organo-compounds, 99
Sulphate of beryllium,
 aqueous systems, 27, 111
 double alkali sulphates, 26
 heat of formation, 28
 preparation, properties and structure, 25
Sulphate routes for opening beryl, 109 *et seq.*
Sulphide, 88
Sulphonate, 33
Sulphuric acid, action on
 beryl glasses, 109
 metal, 3
 phenacite, 34

Telluride, 88

Tritium, formation of, 144

Ultrasonics, absorption by beryllium salt solutions, 11

Valency, see Charge number
Viscosity of
 aqueous salt solutions of beryllium, 7
 molten fluoride of beryllium, 41
Volumetric estimation of beryllium, 120

Willemite, 5, 34

Zenia, colorimetric reagent for beryllium, 121
Zinc, chemical similarity to beryllium, 5
Zinc oxide carboxylates, 5

PRINTED IN THE NETHERLANDS BY
A. W. SIJTHOFF, LEIDEN

546.391　　　　　　　64989
E92c

Everest, David Anthony
The chemistry of beryllium

DATE DUE			

Memorial Library
Mars Hill College
Mars Hill, N. C. 28754